TENTANDA VIA
THE LIBRARY OF
YORK
UNIVERSITY

# The Surface-History of the Earth

Oxford University Press
*London Edinburgh Glasgow Copenhagen*
*New York Toronto Melbourne Cape Town*
*Bombay Calcutta Madras Shanghai*
Humphrey Milford Publisher to the UNIVERSITY

The Drei Zinnen (8,205 ft.; 9,720 ft.; 8,185 ft.) seen from Monte Piano

These extraordinary mountains have been sculptured largely by post-Tertiary denudation out of the horizontally bedded calcareous rocks which prevail in the Eastern Alps.

# The Surface-History of the Earth

BY

JOHN JOLY, Sc.D., F.R.S.

Fellow of Trinity College, Dublin

OXFORD
AT THE CLARENDON PRESS
1925

*Printed in England*
*At the* Oxford University Press
*By John Johnson*
*Printer to the University*

# PREFACE

The Surface History dealt with in this book is based directly upon two great recent advances in our knowledge of the earth's crust: the Radioactivity of the Rocks and Isostasy. The nature of both are herein discussed and explained. Both are recognized to-day by most men of science as *facts*. These two fundamental elements of the surface constitution of the globe lead naturally to certain great events which at long intervals are recurrent in the history of the earth's surface. Such are: (*a*) the slow advance of transgressional seas upon the continents; those seas in which the sandstones and limestones, &c., so common over all parts of the land surface, have been laid down: (*b*) the retrogression of those seas; taking place after many millions of years: and then, finally, (*c*) the rising of the mountain ranges, built mainly of the sediments which were collected in the depths of the former continental waters.

For the genesis of the mountains great lateral as well as vertical forces are required, and these also arise naturally out of those fundamental elements of surface structure: Radioactivity and Isostasy. In the course of this history we recognize eras of tensile crust-stresses changing gradually to eras of compressive stresses. *Inter alia* we find an explanation of the limited scale of the surface relief of the earth; of the causes which have controlled the allocation of land and water upon the globe; and of the major geographical relations of continents and oceans.

With all this to the credit of Isostasy and Radioactivity,

the credibility of the history and of its basis become intrinsically one.

In the course of these studies we have to discard some old ideas and adopt some new ones. The highly heated substratum presented to our mental vision will be new to many. It is, however, no imaginary realm. It, like the other matters with which we are concerned, arises out of our surface observations ; geodetic, seismic, and geologic. This underworld, lurid and tremendous, is very close to us. Yet in serenity we dwell above it ; its cataclysms unseen and unheard. We have to recognize that nothing is more wonderful in a wonderful world than this serenity which surrounds the fragile and perishable organism : permitting its development age after age from lesser to greater things ; culminating in the supreme fact of intelligence.

But the power that redeems the life upon the globe mainly resides in that satellite which hitherto we have regarded as merely ruling the night and raising feeble tides in the ocean : the poet's conception of its mission. In our terrestrial economy, however, it exercises a far more important function. At long intervals in the world's history it becomes the main source of those tides in the liquefied substratum which enable the accumulated radioactive heat of ages to escape harmlessly into the ocean and life to survive from one 'Revolution' to the next.

We acquire, also, new ideas as to our future upon this earth. Our world possesses the gift of rejuvenation ; returning age after age to almost the same starting-point. And Man—if he survives many millions of years yet to come—must see his dominions wrested from him, by uncontrollable forces, to the extent perhaps of more than half his kingdom. But his fate will approach him with almost infinite slowness.

The making of a book like the present is not without difficulties. The ground covered is very wide and to cover it all a much larger volume would be required; for the surface history of the earth involves not only the range of historical geology but also—and what is more difficult to handle—much of physical geology and the related sciences. Hence the many dangers associated with over-concentration attend the writing of a small book like the present.

The ten chapters are written so as to evade these dangers as far as may be, and, it is hoped, so as to be intelligible to a reader possessed of an interest in science and with some knowledge of its rudiments. Matters of detail are relegated to the appendices which close many of the chapters. Some of these appendices are rather more technical than the body of the book, and may be omitted if the reader so desires. At the end of the volume, a lecture delivered before the Geological Society of London in May 1923, and which has not yet been printed *in extenso*, is added. It is hoped that its perusal will enable the reader to obtain a clear retrospective view of the whole matter.

Readers wishing to enter into figures may consult papers in the *Philosophical Magazine* for June and July 1923 ; in which, however, the numerical data used are not in all cases exactly the same as have been accepted in the present volume. A mathematical paper by Mr. J. R. Cotter in the same journal for September 1924, on 'The Escape of Heat from the Earth's Crust', may also be consulted. Some questions arising in the same connexion are treated by Dr. H. H. Poole in the number of the *Philosophical Magazine* for September 1923. A recent account of the whole subject is contained in the Halley Lecture for 1924. On the radioactivity of the rocks, the following may be referred to : *Philosophical Magazine*, October 1912, April 1915, and a recent paper by Dr. J. H. J. Poole and the author, November 1924.

J. JOLY.

Iveagh Geological Laboratory,
Trinity College, Dublin,
*November* 1924.

My thanks are due to Prof. C. Schuchert for permission to reproduce from his *Geology* two stereographic projections of the world ; to M. Armand Colin for the section of the Rhine Valley; to Prof. J. W. Gregory, F.R.S., for leave to reproduce his Map of the Great African Rift Valley; and to the American Geological Survey for the very fine print of the panorama of the Colorado Cañon.

The proofs have been read by Prof. H. H. Dixon, F.R.S., Dr. H. H. Poole, and Dr. L. B. Smyth, to whom I tender my best thanks for much valuable criticism. I also desire to express my thanks to Mr. J. R. Cotter for valuable suggestions.

J. J.

# CONTENTS



2996 B

# Contents

# LIST OF ILLUSTRATIONS

I

# THE EARTH'S SURFACE STRUCTURE

SUCH materials as are accessible to us at the earth's surface reveal a definite stratification according to density. Thus, first the atmosphere ; then the waters resting in lakes upon the rocks or filling the hollows between the continents ; and then the continental rocks averaging a density about 2·7 times that of water. There is evidence that the rocks under the ocean are denser than 2·7. This stratification is, of course, in accordance with the general tendency of physical systems towards stable equilibrium.

If the stratification were perfect we should find the entire earth covered by water which everywhere would be underlaid by the continental rocks. Why this has not come about will enter into our considerations farther on. But what underlies the oceans and continents ? The answer to this question is so important that we must consider it at some length.

In many parts of the earth the heavy rock, basalt, is found to cover vast areas of many thousands of square miles. Much of this surface basalt has been removed by denudation; but what remains is stupendous in amount.

At intervals during the long-past history of the earth great floods of these lavas have reached the surface from the depths beneath. The rock must have been in a very fluid state, for the successive outflows have travelled scores of miles from the fissures which emitted them before congealing. This is the same heavy, black rock-substance with which all are familiar as composing the Giant's Causeway in Northern Ireland ; the Deccan Traps of India; or the Columbia River Lavas of Western North America. Its density is about 3·0. It melts at about 1,150° and a little above this temperature flows like very thin oil.[1] It must have reached the surface at a white heat. The great fissures which emitted these floods opened either on the lower continental levels or upon the sea-floor close to the coasts. We possess no record as to how much may have been poured out on the floor of the ocean.

The well-known Deccan Traps covering much of the north-west of the Peninsula of India,[2] probably covered originally an area of half a million square miles. There still remain over 200,000 square miles overlaid to an average depth of at least half a mile. That is to say, not less than 250,000 cubic miles have probably reached the surface. This great flow occurred towards the end of Cretaceous [3] times or in early Eocene times.

Similar floods broke out on the sea-floor of North-Western Europe extending for some 2,000 miles from the north of Ireland to Franz Josef Land and to an unknown extent westward over the floor of the Atlantic. The mountains of the Inner Hebrides are carved out of their remains. They embraced West Greenland and Iceland and, not improbably, formed a part of yet another vast area of 'plateau' basalts,[4] the 'Siberian Traps' of Northern Russia. They are of Eocene age.

In the Western States of North America the Columbia River basalts cover an area of some 200,000 to 250,000 square miles, their volume being not less than 30,000 cubic miles. These flows reached their maximum in Miocene and Pliocene times. Other enormous floods

[1] Day, Sosman and Hostetter, *Am. Journ. of Sc.*, Jan. 1914.
[2] See the Chart of the World at the end of this volume.
[3] A table of the geological systems is given on p. 88.
[4] So called because of the flat-topped hills to which they give rise.

of similar date occurred in ten areas in the Front Ranges of the Rocky Mountains and along the Pacific Coast. These floods often contain andesites and other lavas, probably the differentiated products of the basalt.

The recent exploration of the great basaltic region of the Parana basin of South America, covering parts of South Brazil, Paraguay, Uruguay, and Northern Argentina shows that this area embraces flows of at least 300,000 square miles extent and a volume of not less than 50,000 cubic miles. A further area of 75,000 square miles, on the eastern margin of the field, is believed to have been formerly covered with basalt. These floods probably took place in Jurassic times.[1] In Patagonia plateau basalts cover many thousands of square miles. Great streams of basaltic lava were poured out in S.E. and N.W. Australia in Middle and later Tertiary times.[2]

While many of these great outflows are of comparative recent date we know that in the earliest times of earth-history similar floods occurred. In very ancient Keweenawan times the Algonkian region around Lake Superior was flooded with similar outflows. In Triassic times—much more recently but still remote compared with the Deccan lavas, &c.—the Palisade basalts flooded vast areas to the west of the Appalachian Mountains.

It is characteristic of these Plateau lavas that all over the earth they are strikingly alike in chemical composition. This remarkable and significant fact has been shown with great clearness by Washington. It plainly supports the view that they are derived from a common reservoir underlying oceans and continents (see Appendix to this Chapter).

In the foregoing pages we have been dealing with the basalt floods arising from the substratum as affording

[1] C. L. Baker, *Journ. Geol.* 31, 1923.
[2] F. R. C. Reed, *Geology of the British Empire*, p. 351.

strong reasons for believing that the substratum is of basaltic character. But we have evidence that the ocean floor itself is everywhere basaltic in character and that it is, in fact, the surface of the general substratum, where it is free from the continental rocks and overlaid by the cooling waters of the ocean. Some of the evidence for this will be referred to in ensuing Chapters. But evidence also arises from the fact that the oceanic islands are mainly built of basalts proceeding from the ocean floor. Again, the chemical resemblance between these surface samples of the substratum is remarkable, as has been shown by a research of Washington's (see Appendix) dealing with a large number of rock samples both from the Atlantic and Pacific Islands.

Now all this strongly supports the view that the great floods originated from a common world-substratum of basaltic lava. Their great volume, their constancy of chemical composition, their wide distribution over the globe, their persistence in time, are all arguments in favour of the existence of the basaltic substratum.

But the evidence goes even further. Petrologists tell us that basalt is not a kind of rock which can originate in the alteration of pre-existing rocks. Daly, who holds this view, says:[1]

'Basalts, diabase, and gabbro must be regarded as primary earth magma.'[2]

And further he writes:

'All visible basaltic eruptives, whether extrusive or intrusive, are best interpreted as due to *abyssal injection* of substratum material along *abyssal fissures* in the crust. Belief in this important principle is impelled by facts and is independent of hypothesis.'[3]

[1] *Igneous Rocks and their Origin*, p. 164 et seq.

[2] Gabbro and diabase are modifications of basaltic lava. See Glossary.

[3] loc. cit., p. 174. Abyssal, i. e. pertaining to the abyss or substratum. So also abyssal fissures open into the substratum.

Von Cotta seems to have been the earliest writer (1858) to urge the view that a continuous basaltic layer lies beneath the earth's acid shell.[1] Daly cites W. L. Green as urging the same view in 1887. He adds:

> 'In 1901 the present writer was independently led to it as the only workable hypothesis for the explanation of the common eruptive sequence illustrated in principle at Mount Ascutney, Vermont. . . . Irrespective of hypothesis, the known distribution of basaltic eruptions both in time and space demand either a very extensive series of subterraneous chambers filled with basaltic material or else a continuous basaltic substratum'; and from the evidence 'so great must be the total area underlain by these imagined compartments of the earth's interior that the whole must form an earth-shell fairly so called'. . . . 'Without some such fundamental assumption the problem of igneous rocks must for ever remain insoluble.'

It is hoped that farther on the reader will recognize that much more even than the solution of the problems associated with the origin of igneous rocks arises out of the existence of a basaltic substratum. That, in fact, the surface history of the earth turns upon its existence and remains inexplicable without it.

The view that beneath the basaltic substratum there is probably a still denser and more basic layer has much in its favour; although the evidence is of quite a different order from that which applies to the existence of the basaltic substratum itself. It is a probability founded on the view that in slowly congealing lavas those first-formed crystals which are heavy must sink to the bottom. Thus, such a mineral as olivine, a silicate of magnesia and iron having a density of 3·3–3·4, would sink in melted basalt. Rocks containing much of this substance and devoid of the lighter silicates (e. g. the felspars) are

[1] i.e. the shell of continental rocks rich in the acid-forming substance silica. In contradistinction rocks containing relatively little silica are termed 'basic'.

'ultra-basic' in composition. Such are 'peridotites', 'magma-basalts', 'dunites', &c. This sort of gravitational differentiation is therefore likely to occur in the substratum and a basal layer of peridotic character to be formed.

On the hypothesis that certain of the Meteorites which reach the surface of the earth represent deep-seated matter ejected by former terrestrial volcanoes they probably give us some idea of the subsurface materials of the earth. The stony meteorites contain, on the average, metallic iron (11 per cent.); ferrous oxide (16 per cent.); silica (39 per cent.); magnesia (23 per cent.); alumina and lime (5 per cent.). Assuming the segregation inwards of the metallic iron, the residue suggests that a peridotic rock may be abundant in the deeper surface layers of the earth. The density of such a rock under normal pressures would be about 3·2.[1]

With the deep interior of the earth we are not here concerned. Suffice it to say that there is no evidence against, and much in favour of, the view that it is largely composed of metallic iron and nickel; both of which enter largely into the composition of metallic meteorites. The magnetic properties of the earth support this view.

The geographical distribution of land and water on the surface of the globe reveals no features which we can identify as determined by its axial rotation. This appears strange, for its axial velocity is great enough to affect the shape of the globe itself. And it is known that a feeble force does in fact urge the continents towards the equator. This force arises out of the high centre of gravity of the continents and the rotational velocity of the earth. It

[1] See Daly, *Igneous Rocks and their Origin*, p. 166.

is very feeble although it has been appealed to as possibly occasioning continental movements in past times.

The present distribution of the land is clearly not equatorial. Much of it is extended north and south. We may say that the continental 'grain' of the earth is meridional. The Americas and Africa are great instances. Further, the entire region from Siam to Tasmania, which includes the shallow, land-girt waters and islands interposed between Australia and China, forms one great continental peninsula directed meridionally towards the south. The result is that the equator traverses the ocean and not the land for by far the greater part of its length.

Broadly speaking, *the dominant geographical feature of the globe is the relation of the continents to the oceans; the latter dividing up the land and extending from pole to pole.*

And we should observe that while the land seems to shun a belt-like distribution around the earth, either along or parallel with the equator, so also it shuns aggregation at either pole. The North Pole is girt round by a deep sea. Antarctica (but little known) is penetrated at least to the 80th parallel by the ocean.

All these geographical features we must note. The causes from which they arise will be later considered.

The mountain-grain of the earth is conspicuously oriented in two directions, approximately perpendicular. The mountains of North and South America extend approximately north and south, and so do those of Australia and New Zealand, and the ranges of eastern Asia, as well as the Ural Mountains and some in Europe and Africa. The chief mountain ranges of Eurasia trend approximately east and west. And these relations have been, for the greater part, maintained from the very remote past. The amount of orogenesis (mountain building) apparent at the present time as having been effected in each of these two directions does not seem to differ greatly.

The relations of terrestrial mountains to the oceans is an extraordinary one. In many cases they border the oceans just where we should not expect them to be. They rise like a great wall around the Pacific—attaining their greatest altitude on its eastern coasts. And these ranges first came into existence in very remote times and have shed, in process of denudation, probably far more than their present bulk into the rivers and seas. But still their successive generations took their origin parallel with the ocean boundaries.

Around the Pacific the mountains are largely volcanic. Not that they are mainly composed of lavas, but they are topped with lava cones often many thousands of feet high.

Around the Atlantic, mountain development is comparatively inconspicuous. The Appalachians are mountains of old formation to which recent events have contributed vertical movements only. If produced northeastward, it is said, they appear to form a continuation of certain old mountains of like age in Western Europe—the Armorican mountains. The other mountains around the Atlantic do not range parallel with its shores or rise from deep water. Nor are they volcanic. They are relatively small and are often cut obliquely by the coasts.

The greatest mountains of the earth, the Himalayas, have only arisen in recent times. Except for the interposition of the Peninsula of India, once a part of Gondwanaland,[1] the ranges of the Himalayas confront the greatest ocean stretch upon the earth. Measure on a terrestrial globe, the span of ocean lapping round the world which confronts this region of high mountains,

[1] See Chapter VIII.

and you will find it to be more than 13,000 miles. The opposite shore of this stretch of ocean is the Pacific coast of Mexico.

That 'the greatest mountains confront the widest oceans' was pointed out by Dana long ago. It applies as we have seen to most of the ranges of the earth. Some connexion between the development of the mountain range and the ocean must surely exist. What is the nature of that connexion? In subsequent chapters we shall consider the answer. But a superficial feature which must enter frequently into our considerations should be referred to here; the geological examination of the great mountain ranges reveals the fact that the materials of which the mountains are composed appear to have been in some past time exposed to enormous forces coming from the direction of the ocean. There seems no doubt that the forces associated with the mountain structures along Western America proceeded from the Pacific.

Ransome[1] says of this view:

'It has been a rather widely held tenet among Geologists that most of the great mountain ranges of the globe have been formed by thrusts from the nearest ocean basin, the classic example being the Appalachians. It has also been believed by many that the Laramide system is essentially Pacific in origin and that the deformation which gave it birth was the consequence of interaction between the Pacific Basin and the continental margin.'

Willis states:[2]

'The folding which occurred during late Cretaceous and Eocene throughout the Rocky Mountain province, even as far east as the Front Range of Colorado, is to be attributed either to Pacific thrust or to pressure exerted from the region of the Great Plains.'

Having considered the inadequacy of the latter as in any

[1] *Problems of American Geology*, pp. 364–5.
[2] *Bull. Geol. Soc. America*, vol. xviii, p. 406, 1907.

way a potential source of force, Willis concludes by stating his conviction that it is the Pacific which was

'the source of the orogenic activity which has produced the Cordillera from Alaska to Cape Horn'.

Keith, speaking of the orogenetic movements which developed the Appalachians, concludes that they were directed from the south-east.[1]

It would be easy to quote many distinguished geologists as expressing similar views. The great Eastern Eurasian Chains of fold-mountains are very generally regarded as having been developed by forces proceeding from the south or south-east, that is, from the great stretch of ocean to which we have already referred. And this applies not only to the recent chains but to the far more ancient Armorican Mountains. Hobbs, in a study of the curved ranges of Eastern Asia (including the Festoon Islands), gives good reasons for the belief that the mountain-building force proceeded from the ocean, and indeed cites evidence that the ocean floor is still active and is slowly tilting up the eastern sides of the Japanese Islands.[2]

We perceive, therefore, from the structural features of the earth's surface that great *compressive* forces have affected the continents. But not the continents only. The ocean floor also shows signs of similar effects. Gentle anticlines where the floor has been bowed upwards traverse it for many hundreds of miles. One such traverses the Atlantic from extreme south to extreme north. Great 'deeps' or troughs where the floor has been bowed downward many thousands of fathoms are found to exist here and there in the floor of the ocean and often following the curvature of the land.[3] They are often

[1] *Bull. Geol. Soc. America*, vol. xxxiv, p. 338.

[2] 'The Asiatic Arcs,' *Bull. Geol. Soc. America*, vol. xxxiv, 1923.

[3] See the Chart of the World at end of this volume.

associated with continued seismic movements—earthquakes and earth-tremors—as well as with volcanicity. The great Tuscarora Deep bordering the restless Japanese Islands is a conspicuous example.

But the earth's surface everywhere exhibits evidence also of past *tensional* forces. Perhaps the most conspicuous example of this is the great African Rift Valley. (Plate II, p. 27.) The long narrow lakes of eastern Africa lie in the down-sunken floor of this great rift. Two approximately parallel rifts developed in Oligocene times in the continent from south to north. There was first an upward bowing of the land. After the rifts or faults had formed, penetrating down to the very base of the continent, the narrow tract between sank down 5,000 feet or more. How this came about we shall see later. The Red Sea fills a great spreading-apart of the two rifts. The Valley of the Jordan is formed in the same manner. Volcanic phenomena broke out along its length and basaltic lavas were ejected. It is four thousand miles long, extending over one-sixth of the circumference of the globe. Suess, and others who have studied it, recognize its tensile origin. Suess says:

> 'A process which finds expression over more than 52 degrees of latitude must have origin in the structure of the planet itself. For this great region we are led to assume *the existence of tensions in the outer crust of the earth which have acted in a direction perpendicular to that of the fissures* and in this case, as it happens, perpendicular to the Meridian.'

Lesser rift valleys with a meridional trend occur elsewhere upon the globe. Meridional rifting on a great scale is conspicuous in South-Eastern Australia.[1] The Christiania Fiord and the valley of the Rhine are rift valleys revealing tensional effects. In the case of the Rhine valley the floor is an inverted arch of younger rocks

[1] Reed, loc. cit., p. 351.

let down by normal faulting among more ancient strata from the surface of which denudation has removed the younger rocks.[1] (Fig. 1, p. 28.)

Normal faulting on a great scale is responsible for the formation of the great Rift Valley; and normal faulting in general results from the action of tensile forces, as Sollas long ago pointed out. The opposite sides of the sloping fracture are displaced in such a manner as to lead to extension of the crust. Reversed faulting arises, on the other hand, from compressional forces; the opposite sides of the fracture being so displaced as to involve contraction of the crust. The miner recognizes the presence of such faults everywhere ; extending deep down ; tensional as well as compressional. Daubrée and Kayser insist on the frequency of tensile fissures in the continental crust. Dutton cites cases where mountain-forming blocks have been let down by the separation of supporting blocks drawn apart by tensile forces. Doubtless, those fissures through which great volumes of basalt are brought up from the substratum, whether formed in the sea-floor or upon the land, are due to tensile forces. What is, probably, further evidence of tensile effects in the ocean floor is the alignment of the Pacific volcanic islands. An ocean floor stretched from shore to shore would be rifted in the directions shown by the island chains, for, in fact, they extend perpendicularly to its greatest span, bearing north-west to south-east. Rifts so formed would lead to volcanism actuated from a substratum of fluid basalt beneath the ocean floor.[2] (Fig. 2, p. 29.)

He who has visited any of the great mountain regions of the earth is impressed by the greatness of the forces

[1] Haug, *Traité de Géologie*, i, p. 242. For definition of normal and reversed faulting see Glossary.

[2] Dana's *Manual of Geology*, 3rd ed., p. 32.

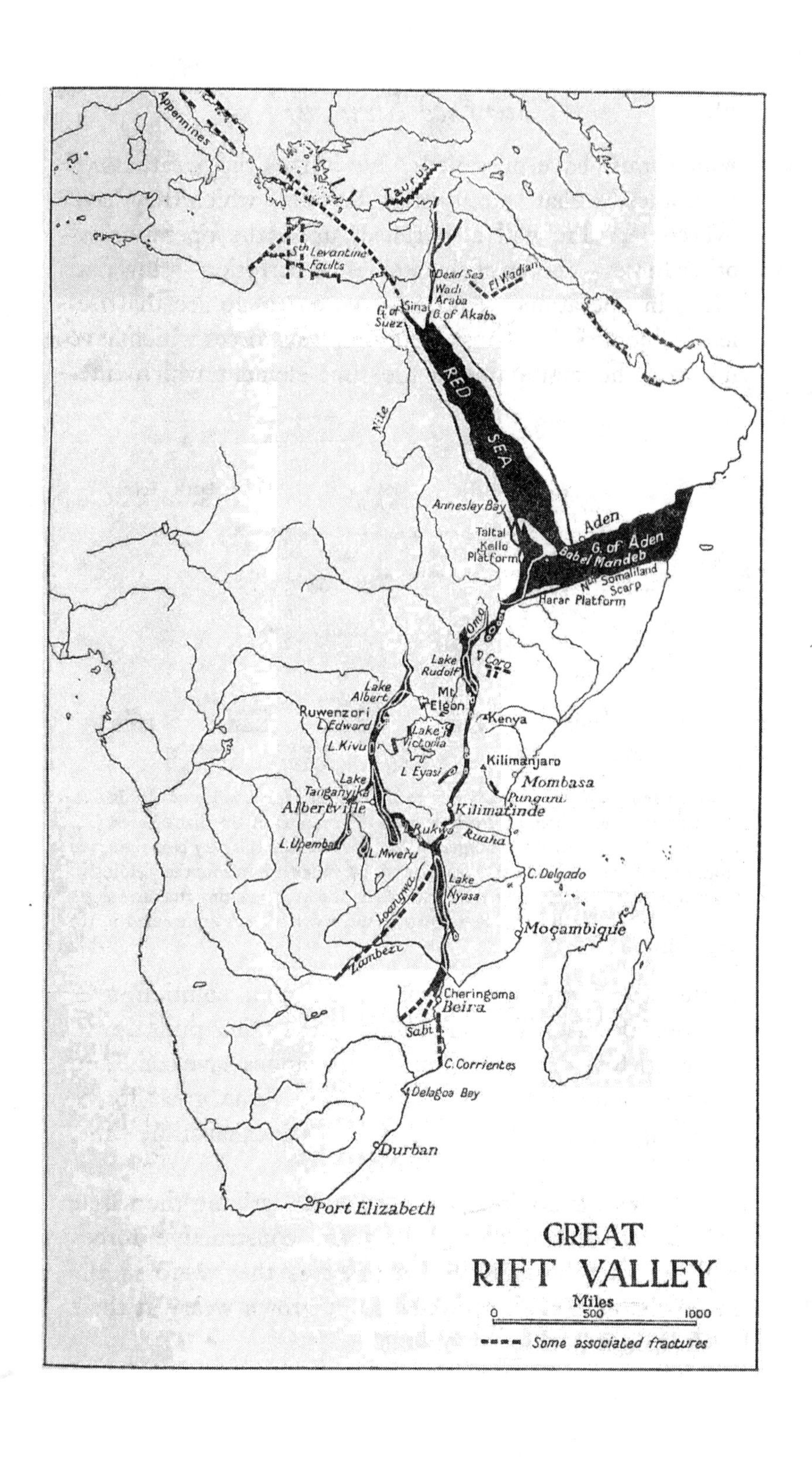
Appennines
Taurus
Sth Levantine Faults
Dead Sea
Wadi Araba
El Wadian
Sinai
G. of Suez
G. of Akaba
RED SEA
Nile
Annesley Bay
Taltal Kello Platform
Aden
G. of Aden
Babel Mandeb
Nth Somaliland Scarp
Harar Platform
Omo
Coro
Lake Rudolf
Lake Albert
Mt. Elgon
Ruwenzori
L. Edward
Kenya
Lake Victoria
L. Kivu
L. Eyasi
Kilimanjaro
Mombasa
Lake Tanganyika
Pangani
Albertville
Kilimatinde
Rukwa
Ruaha
L. Upemba
L. Mweru
C. Delgado
Lake Nyasa
Loangwa
Moçambique
Zambezi
Cheringoma
Beira
Sabi
C. Corrientes
Delagoa Bay
Durban
Port Elizabeth
GREAT RIFT VALLEY
Miles
0 500 1000
Some associated fractures

which must have prevailed. He judges the greatness of the forces by that 'strength of the hills' which they have overcome. He will also reflect upon the operation of other forces—forces of the feeblest description—slow and silent in operation. And here he comes to see that not less influential than those overwhelming forces which have uplifted the mountains is the time-element which inte-

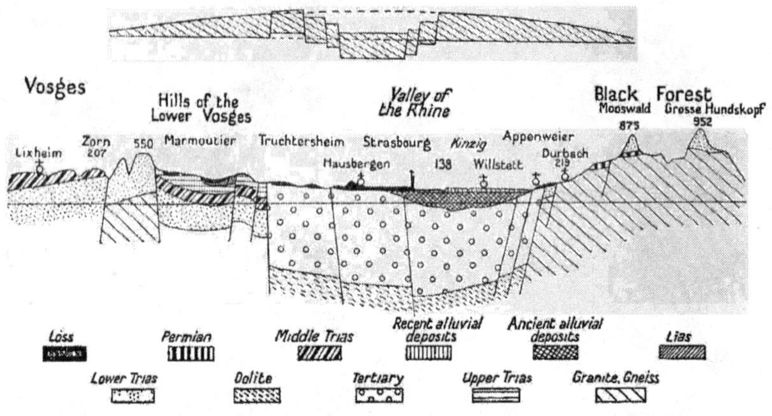


Fig. 1 shows in cross-section (W. to E.) the foundered valley of the Rhine. Tertiary rocks have been let down between the granites of the Black Forest (E.) and the Triassic rocks of the Vosges (W.). They have in this way been preserved from denudation. Plainly the structure of this valley arises from the yielding of the continental crust to tensile forces exerted E. and W.; creating fractures along which the crust has been faulted down in successive steps as represented in the upper diagram.

grates the feeble forces of friction and solution over geological eras. The one uplifts; the other pulls down. In the history of the earth mountain ranges have come and gone; come and gone many times. The great forces uplifting; the feeble ones, aided by inexhaustible time, pulling down.

But there is obviously a mystery underlying the whole matter. Whence come the great constructive forces which to-day seem to be as great as they were in the most remote past? They have not grown weary at their work; for the existing mountain ranges—of recent date as we shall see—are equal to, if they do not exceed, those of former ages.

The mystery deepens when we are told that these great folding forces proceed from the ocean and that their magnitudes are measured by the ocean-span.

We find also that crushing is not the only form of

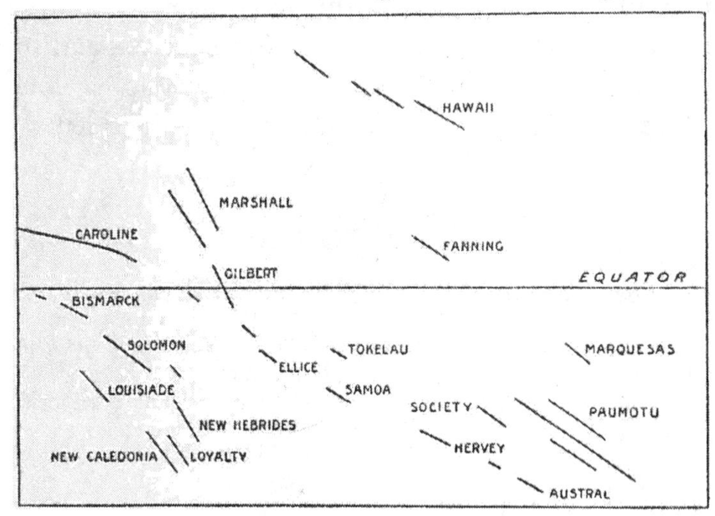


Fig. 2. Trend lines of the Islands in the Pacific.
After J. D. Dana. Scale 1 : 100,000,000

stress which has racked the continents. Irresistible tensions have also prevailed in the past and we have presented to us the spectacle of one of the greatest of the continents rent from end to end by these tensile forces. We ask in turn whence come these great tensional forces.

But these facts do not exhaust the element of mystery prevailing over the surface features of the earth. Our cities are raised on rocks which formerly were far beneath the surface of the ocean. Right across the great continents stretch the floors of ancient seas. And geological research tells us that not now for the first time have these continental regions risen from the ocean to receive the light of the sun. There were repeated submergences and repeated resurrections.

Finally, and most remarkable of all, an orderly sequence has prevailed in these great events; the entire surface history of the earth being, as it were, laid out according to a succession of these events, sundered by enormous intervals of geological time.

We shall find later that the phenomena of the resurrection of the land and of the mountains are physically connected—spring from a common source—and in due course both land and mountain range find a common grave in the ocean waters rising over the continents.

## APPENDIX TO CHAPTER I

### *The Chemical Resemblance of the Plateau Basalts*

Washington has recently published the results of his chemical analyses of these rocks, which are here reproduced.

It will be seen that the likeness in chemical composition is remarkable.

| | *Deccan* 11 | *Oregonian* 6 | *Thulean* 33 | *Patagonia* 1 | *New Jersey* 8 |
|---|---|---|---|---|---|
| $SiO_2$ | 50·61 | 49·98 | 47·46 | 50·74 | 50·66 |
| $Al_2O_3$ | 13·58 | 13·74 | 13·89 | 12·60 | 14·28 |
| $Fe_2O_3$ | 3·19 | 2·37 | 3·58 | 4·78 | 3·41 |
| $FeO$ | 9·92 | 11·60 | 9·38 | 7·25 | 8·58 |
| $MgO$ | 5·46 | 4·73 | 6·79 | 9·00 | 6·92 |
| $CaO$ | 9·45 | 8·21 | 9·83 | 8·90 | 8·60 |
| $Na_2O$ | 2·60 | 2·92 | 2·90 | 2·59 | 2·92 |
| $K_2O$ | 0·72 | 1·29 | 1·01 | 0·72 | 0·72 |
| $H_2O$ | 2·13 | 1·22 | 1·48 | — | 2·28 |
| $TiO_2$ | 1·91 | 2·87 | 2·71 | 2·73 | 1·30 |
| $P_2O_5$ | 0·39 | 0·78 | 0·43 | 0·37 | 0·17 |
| $MnO_2$ | 0·16 | 0·24 | 0·22 | 0·31 | 0·12 |
| | 100·12 | 99·95 | 99·78 | 99·99 | 99·96 |

'Oregonian' signifies basalts from the Oregon region of W. America. 'Thulean' signifies basalts from the Hebridean region of North-West Europe.

It is of interest to note that compared with other basalts—such as are ejected by volcanoes, and which probably come from sources nearer to the surface—' the chief difference is in the much higher amounts of iron oxides with ferrous oxide greatly predominating over ferric oxide '. There is also a higher percentage of titanium in the plateau basalts (Washington).[1]

### *The Atlantic and Pacific Basalts*

As the result of seventy-two chemical analyses of rocks derived from the floor of the Atlantic Ocean and fifty-six analyses of rocks from the Pacific floor, Washington finds the means of the Atlantic and Pacific basalts to be as follows :

| | *Floor of the Atlantic* | *Floor of the Pacific* |
|---|---|---|
| $SiO_2$ | 50·63 | 50·06 |
| $Al_2O_3$ | 15·82 | 15·51 |
| $Fe_2O_3$ | 4·44 | 3·88 |
| FeO | 5·73 | 6·23 |
| MgO | 5·79 | 6·62 |
| CaO | 7·36 | 7·99 |
| $Na_2O$ | 4·27 | 4·00 |
| $K_2O$ | 2·31 | 2·10 |
| $H_2O$ | 1·47 | 1·16 |
| $TiO_2$ | 1·63 | 1·96 |
| $P_2O_5$ | 0·43 | 0·25 |
| $MnO_2$ | 0·04 | 0·15 |
| residue | 0·07 | 0·08 |
| | 100·05 | 100·00 |
| Sp. gr. | 2·85 | 2·89 |

These results agree generally with the composition of a basalt possessing rather high silica and alkali percentages. The specific gravity is below that of the more deep-seated varieties of these magmas (dolerites and gabbros).[2]

### *Magmatic Differentiation*

To enter at all fully into this subject the reader must possess a fairly intimate acquaintance with the nature and structure of many rock clans and rock-species. But an idea of the principles involved and the results arising therefrom may be easily acquired. The view

[1] H. S. Washington, *Bull. Geol. Soc. of America*, vol. xxiii.

[2] Cited by Loewinson-Lessing, *Bull. Soc. Geol. de France*, t. xxiii, 1923, p. 142.

advocated by Daly, and held regarding many aspects of it by other eminent petrologists, is, briefly, that granites are derived 'as the final acid pole of the earth's primitive differentiation whereby the acid shell and the basaltic substratum became separated'.[1] 'This theoretical conclusion seems to be amply supported by the facts of nature.' The principle involved is that under gravity the heavier constituents, crystallizing out, sink and the lighter rise in the very slowly cooling magma. Very commonly a graded series of rocks passing upwards into granites may in this way be formed.

Similar views apply to the formation of andesites out of a basaltic magma. Andesites contain less iron and more silica than basalts. As iron is a heavy constituent and silica a light one the andesites are of lower specific gravity than the basalts. It is believed that even if mingled in the fluid state there may be light and heavy liquid fractions separated under gravity and that this process of 'gravitative differentiation' may be assisted by a certain limited miscibility of the molten magmas just as cream cannot be kept mixed with milk but separates out.[2] In support of the 'obvious and long recognized hypothesis that augite andesite is a differentiate from basalt' Daly cites the experimental work of Doelter and others as leading to the unavoidable conclusion that olivine, augite, and magnetite must settle out of basalt kept in a fluid state. 'We have then to expect in nature a continuously graded series of lavas from pure olivine basalt, through olivine-free basalt to those phases of the mother liquor which must approximate to a basic augite andesite and then an acid augite andesite.' 'Many peridotites, the picrites, limburgites (magma basalts), and abnormal olivinitic basalts' (i. e. 'ultra-basic' rocks) 'are, on this view, the rocks derived from the fractional crystallization of olivine basalt, while augite andesite or allied types represent the other pole of the differentiation.'[3] Many other considerations are advanced.

We infer from this that the andesites ejected from the great mountain ranges from which they derive their name are not primary rocks but are derived from the great basaltic substratum lying far beneath.

[1] Daly, loc. cit., p. 361. [2] loc. cit., p. 229. [3] loc. cit., p. 377.

## II

## ISOSTASY

Some seventy years ago, in carrying out the trigonometric survey of Northern India, it was ascertained by triangulation that the difference in latitude between Kalianpur and Kaliana was 5° 23′ 42·294″; whereas the astronomical observation in which a plumb-line had been used to define the zenith, showed a difference of 5° 23′ 37·058″. That is to say 5·236″ less than was ascertained by triangulation. The more northern of these stations, Kaliana, was some sixty miles south of the Great Massif of the Himalayas.

It was inferred that the attraction of the mountain ranges had affected the plumb-line—thereby falsifying the result. Archdeacon Pratt thereupon made by direct calculation an estimate of what the error ought to have been, assuming the density of the Himalayan Mountains was 2·75 times that of water. He found that the error should have been considerably *more* than what had been observed. It ought to have been 15·885″ according to his estimate of the attractive force of the rocks that build up the great mountain ranges. And the question arose: why did the mountains fail to exert their full attractive effect upon the plumb-line?

This important result led Sir George Airy—then Astronomer Royal—to suggest an explanation which, in its results, has revolutionized our views as to the surface structure of the earth.

He first shows that mountains so great must be supported from beneath, and could not be merely resting upon a uniform crust supported by a dense viscid lava, even if that crust were 100 miles thick. He then asks :

> 'What can be the nature of its support ? I conceive there can be no other support than that arising from the downward projection of a portion of the earth's light crust into the dense lava : the horizontal extent of that projection corresponding widely with the horizontal extent of the table-land.'

(That is a hypothetical table-land calculated to represent the mass of the Himalayas.) From this theory Airy explains the deficient disturbance of the plumb-line as follows. He says :

> 'It will be remarked that the disturbance depends upon two actions : the positive attraction produced by the elevated table-land, and the diminution of attraction produced by the substitution of a certain volume of light crust (in the lower projection) for heavy lava. The diminution of attractive matter below, produced by the substitution of light-crust for heavy lava, will be sensibly equal to the increase of attractive matter above.'

In other words, the mountains are *floating* in the dense substratum.

Exactly the same result would be obtained if we set up a plumb-line near an ice-floe or iceberg floating in the sea. There would be, at some distance away, practically no distinctive effect on the plumb-line due to the emergent ice. For a great mass of ice, having a density less than that of fresh water, projects downwards sufficiently far to displace a mass of the heavier salt water precisely equal to the weight of the floe or berg, this being an elementary condition of flotation. Hence the total mass attracting the plumb-line in the direction of the berg would be no greater than that due to the ocean which attracts it in the opposite direction. There will be no resultant deflexion. Similarly an observation upon the Himalayas at a sufficient distance would show no distinctive effect due to the visible mountains.

The now well-known and much-investigated theory

of isostasy (so named by Dutton) originated in Airy's explanation.[1]

This theory involves the view that the lighter continental crust floats upon a universal substratum of heavier materials. It follows that the thickness or depth of the continental layer is much greater than is visible to our observation. For supposing that we—on the strong evidence given in Chapter I—assign to the substratum the density of deep-seated basaltic rocks (gabbro)—i. e. 3·0—and if the average continental rocks (mainly granites, gneiss, and such-like highly siliceous rocks) possess an average density of 2·67, it will follow that the immersed part must extend downwards about eight times the height of the emergent part; or, in nautical terms, the 'draught' is eight times the 'freeboard'. To see this, consider that a cubic centimetre of basalt weighs 0·33 grm. more than a cubic centimetre of granite. In other words, the displacement of one c.cm. of basalt by one c.cm. of granite will float a weight of 0·33 grm. Hence it will require the submergence of 2·67 ÷ 0·33 = 8·09 c.cm. of granite—say shaped as a vertical column—to support one c.cm. of emergent granite. The emergent volume therefore bears to the submerged volume approximately the ratio 1 : 8.

The mean height of the continents over sea-level is stated to be 0·82 kilometre.[2] The mean depth of the ocean is 3·8 km. (Murray). Adding these heights, we have it that the mean continental surface is 4·62 km. over the sea-floor. Let us take it that the sea-floor *is* the surface of the substratum. Then we have it that the continents project 4·62 km. above the surface of the substratum.

[1] For a good account of the matter see O. Fisher's *Physics of the Earth's Crust*, p. 194 et seq.

[2] The most recent estimate (1912) due to Wagner, quoted by Supan, *Grundzüge der physischen Erdkunde*, 6th ed., p. 52.

We cannot, however, calculate the submerged depth without making an allowance for the buoyancy effect of the oceans. The allowance for this shows that the 3·8 km. of continental rock submerged in the sea is effectively reduced to a stratum 2·4 km. deep. Hence the effective emergence of the continents above the surface of the substratum is $2{\cdot}4 + {\cdot}82 = 3{\cdot}22$ km. If we multiply by 8·09 we obtain the depth submerged in the substratum: 26·05 km. And the total average depth of the continents is $26{\cdot}05 + 4{\cdot}62 = 30{\cdot}67$ km.

We shall have occasion to make other estimates of this kind. The reader must bear in mind that the results obtained must be regarded as approximations only. We cannot claim to know any of the data accurately.

Meanwhile a sketch drawn roughly to scale will enable us to realize what our estimates involve (Fig. 3). We are looking at a vertical section of—suppose—the Himalayas. To the left the normal continental crust extends. At the cross we have set up a plumb-line. Now recall Airy's explanation of the effect of the range upon the plumb-line. We shall introduce here a very useful word. The effect of the mountains upon the plumb-line is 'compensated' by the great bulk of light material which, beneath them, has taken the place of the heavy basalt.

Similar conditions prevail all over the surface of the globe ; not on a small local scale, but where great areas, great mountain ranges, or great expanses of water are considered. The law expressing this fact is known as Isostasy.

To understand what isostasy involves we should attend to a very fundamental law of flotation. When any medium supports other substances in flotation the mass beneath unit area remains everywhere alike. Take the case of an ice-floe in the sea. And, to make our reasoning more precise, suppose we are considering the total mass per unit area as measured down to an imaginary datum plane—say one mile beneath the surface. Then the law referred to involves that the mass is the same beneath any superficial unit area, whether this is taken over the ice-floe or over the bare surface of the sea. The truth of this law depends, of course, upon the well-known principle of Archimedes that a floating body must displace a mass of the sustaining fluid precisely equal to its own mass.

Similarly with respect to the earth's surface. If isostasy everywhere prevail—or, as would be said, if

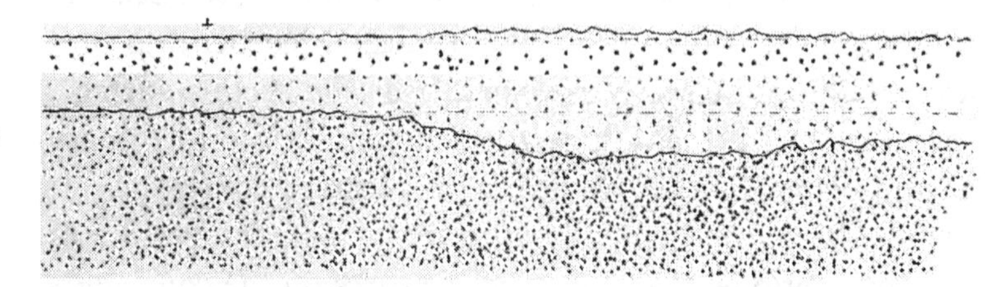

Fig. 3. Diagram of the Himalayas and their isostatic compensations. The cross marks position of plumb-line observations

everywhere there were 'isostatic equilibrium', then the mass measured down to any imaginary datum plane extending parallel with the earth's surface and at a sufficient depth in the substratum, is the same per unit area taken over the earth's surface.

The surface of the earth here spoken of is the mean level of the ocean or the surface of a tideless sea. This is known as the *geoid* and is the surface to which the direction of gravitation is everywhere perpendicular. Hence it extends everywhere perpendicularly to the direction of the plumb-line.

Before entering further into the evidence for isostasy we must clearly understand what it involves.

The continents are floating. Each continent displaces

a mass of basalt in the substratum equal to its own mass. The ocean simply rests upon the surface of the substratum. The ocean floor *is*, therefore, this surface. Each continent is 'compensated' as a whole. The ocean is 'compensated' in the sense that its deficiency of mass is made good by the rise of the substratum underlying it.

An important part of the investigations carried out upon isostasy is directed to the question how far compensation is local or how far it is regional. It cannot be said that any clear decision has been arrived at upon this point. There seems no doubt that isostasy prevails over the earth as a whole; that is, that continents and oceans are in isostatic equilibrium. Which means that beneath unit areas, taken sufficiently extended over the earth's surface, there is equal mass measured down to an imaginary datum plane whether we deal with ocean or continent. It seems also to be generally admitted, upon the results of the observations, that the larger features on the continents are locally compensated, e. g. the Himalayas; the Rockies; the Appalachians; the Great Plains. Some of the best authorities, e. g. Hayford and Bowie, consider that local compensation extends to areas of about one square degree.

How far local compensation may extend in the substratum beyond the horizontal extension of the surface feature which it supports is not known, or whether it so extends at all. As to the smaller topographical features, it is evident that the strength and stiffness of the continental crust would distribute their weight, and local compensation could not exist. So also it is certain that individual peaks of the mountain ranges give no clue to the depth of compensation. It will appear later that their individual compensations in many cases would not be stable. The compensation of a mountainous area must be determined by the average height, or the height of an equivalent plateau.

Since the discovery of Pratt and Airy a very great amount of work has been expended on the investigation of isostasy. It may be tested either by the plumb-line or by the pendulum.

The plumb-line gives the direction of the resultant gravitational force at any place. But it can tell us a good deal about isostasy; and has led, as we have just seen, to its discovery.

Hayford early in the present century carried out geodetic work in the United States directed to the investigation of 'the Figure of the Earth and Isostasy' which has become historic. His earlier work was restricted to the use of the plumb-line.

The work consisted of determining the geodetic latitude and longitude of many hundreds of stations scattered over thirty-three of the States. These determinations are effected by triangulation—that is, by direct mensuration over the earth's surface—connecting the stations with one reference or base station, the latitude and longitude of which have been very carefully determined.

Now the astronomic latitude of a place is defined as the angle between the direction of a plumb-line at the place and the plane of the equator. Hence this definition of latitude assumes that the plumb-line points to the zenith. Upon that assumption the plumb-line can be used to find the latitude.

Hayford tabulates the geodetic latitude and longitude and the astronomic latitude. Subtracting the geodetic from the astronomic latitude he arrives at a quantity—the 'residual'—which may be plus or minus, and is never more than a very few seconds of arc. It is that component of the deflexion of the vertical (at the particular station) which lies in the meridian; deflexion meaning the angle between the actual line of gravity at the station (i. e. as given by the plumb-line) and the line which is normal to a certain spheroid, called the Clarke spheroid, which most nearly represents the surface of the earth and which is used in the geodetic determinations. If the sign is plus, the zenith defined by the actual line of gravity is farther north on the celestial sphere than the zenith defined by the line normal to the Clarke spheroid. In a similar manner the deflexions in a vertical circle passing through the east and west points, i.e. in the prime vertical, are determined; the plus sign then signifying a westerly deflexion of the zenith.

As an instance we find that for Salt Lake City the geodetic latitude is 40° 46′ 12·38″, and the astronomic is 40° 46′ 03·36″; giving a residual of —09·02″. The plumb-line, therefore, points to the south of the geodetic zenith by this amount.

As contributory to the evaluation of these deflexions of the plumb-line, firstly all topographic irregularities have to be taken into account. This is just what Pratt in part did in the momentous calculations which disclosed the whole secret. It is evident that a hill to the north of a station will attract the plumb-line so as to produce a southward displacement of the zenith at the station. With the expenditure of immense labour the attractive effects of topographic features within 4,126 kilometres were calculated.

When the computed topographic deflexions of the plumb-line all over the States were compared with the observed deflexions of the vertical, the latter were found to be much smaller than the former. The logical conclusion reached by Hayford was 'that some influence must be in operation which produces an incomplete counterbalancing of the deflexions produced by the topography, leaving much smaller deflexions in the same direction'—which is the same idea that presented itself in explanation of the Himalayan observation. Hayford concludes

'that there must be some general law of distribution of sub-surface densities which fixes a relation between sub-surface densities and the surface elevations such as to bring about an incomplete balancing of deflexions produced by topography on the one hand against deflexions produced by variations in sub-surface densities on the other hand. The theory of isostasy postulates precisely such a relation between sub-surface densities and surface elevations.'

In his earlier work Hayford does not deal with observations on the force of gravity. Later, in conjunction with Bowie, much valuable work in this direction was accomplished.

Of the many methods suggested for determining the force of gravity none seem to be so reliable as the pendulum. The vibrating pendulum is essentially a falling body, the acceleration of which is utilized to cause the body to rise again to near its starting-point. The fall is repeated twice in each complete vibration. The time of completing a vibration therefore depends on the gravitational force. It will vibrate faster at the poles than at the equator, the force of gravity being greater at the poles.

In general the pendulum is let swing for about forty minutes under its own inertia—corrections are made for its expansion with change of temperature, for the density of the atmosphere (unless it is enclosed in an exhausted space), and for vibrations set up in its support. Calculations of the force of gravity are based upon its time of vibration as deduced by accurate comparison with the rate of swing of a standard pendulum. This may be kept at a base station at which, by a special investigation, the force of gravity is accurately known.

Determinations of the force of gravity afford a direct means of investigating the perfection of isostatic conditions in any locality on the continental surface. But before the observations can be applied to this purpose many sources of disturbance have to be calculated out. The latitude and longitude, the altitude above sea level, local and distant disturbing topographic features extending even to the Antipodes, have to be taken into account. A computed value of the force of gravity for the particular station is arrived at and this is compared with the observed value. If there is a difference this difference is known as 'the gravity anomaly'. The gravity anomaly gives the local departure from perfect isostasy. It is generally small. The result is expressed in dynes.[1] Now the force of gravity over the earth varies around 980 dynes. That is, a weight let fall at the earth's surface acquires an acceleration of about 980 cm. per second per second. The results of the pendulum observations on the force of gravity at any locality may afford a number such as 978·970 (e. g. for Key West, Florida). The computed value for this station is 978·957. The difference is 0·013 dynes, and this would be known as the gravity anomaly at this station.[2]

Hayford and Bowie in their calculations do not proceed according to the assumption of flotation by displacement.

[1] A dyne is the force which confers an acceleration of 1 cm. per second on a mass of 1 grm. See Glossary.

[2] The computed value is derived from the following quantities: Latitude of Key West 24° 36·6′: longt. 81° 48·4′; giving theoretical gravity 978·922. Elevation over sea 1 metre; correction for elevation = 0·000. Correction for topography and compensation 0·035. Hence computed gravity (978·922 + 0·035 = 978·957) as above.

They assume a surface structure more amenable to mathematical treatment. They assume, indeed, the fundamental law of flotation (that there is equal mass beneath equal areas) which we have already discussed, but

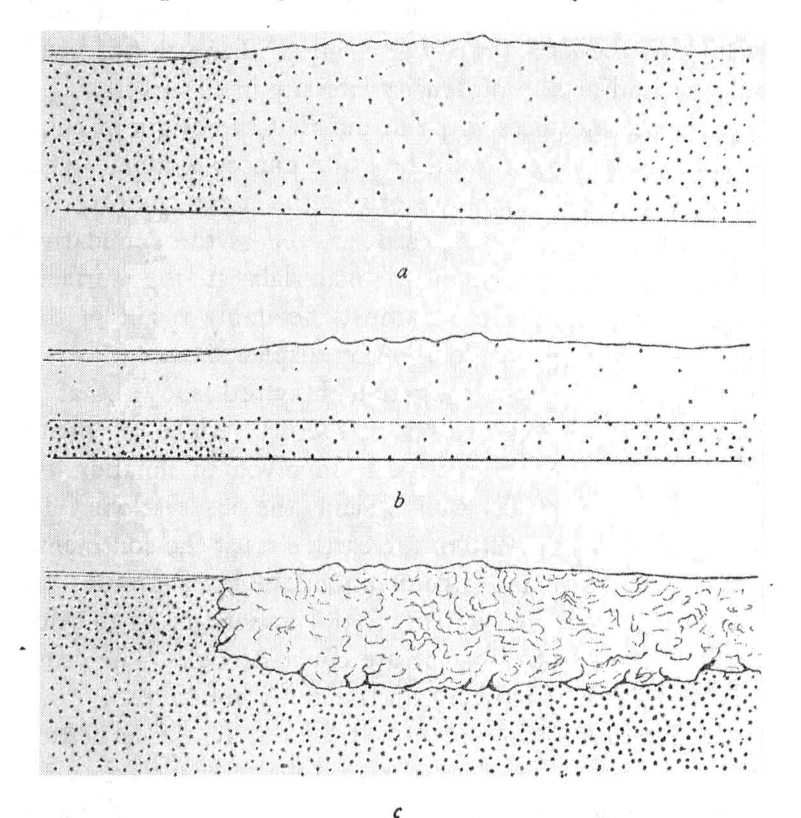


Fig. 4. Diagram showing: (*a*) Hayford's theoretical crust, compensations being uniformly distributed down to a certain plane. (*b*) Hayford's alternative crust with compensations restricted to a layer 10 miles deep beneath continents and oceans. (*c*) The probable reality.

arrive at it upon a hypothetical distribution of density in the outer crust which is illustrated in Fig. 4 *a*. The density of the continental crust is assumed to vary with its surface elevation above a certain plane or level of complete compensation. Beneath mountains the crust is light; beneath the ocean it is heavy, and so on. In this manner there is equal mass beneath equal horizontal areas taken over the earth. The depth at which the assumed plane of complete compensation exists is investigated by trial calculations directed to find what level will give the least gravity anomalies. Beneath this level there is uniformity of density laterally in all directions.

It seems needless to point out that the origin of such a continental crust would be quite unaccountable. And even if we could explain its origin, the lateral compressive forces to which it is exposed, as well as the denudative and volcanic movement of materials at the surface, must soon disturb it and must inevitably result in the natural conditions of flotation by displacement.

By a separate investigation Hayford shows that a second hypothetical surface structure, which in many respects is identical with that involved in flotation by displacement, equally well explains the observations. In this assumed structure of the earth's crust the continents are supposed to be directly underlain by a stratum 10 miles thick within which all the variations of density which give rise to isostasy are concentrated. The crust above is assumed to be 25, and again 27, miles deep. Both agreed with the observations—the latter a little the better. This assumed distribution of density obviously is very similar to flotation by displacement. The lighter parts of the 10-mile layer represent the compensations penetrating downwards into the heavy magma; the denser parts the invading areas of the heavier medium. It evidently corresponds with the density distribution of the third diagram.

Further on it will be seen that a fatal objection besets the crust primarily assumed by Hayford. He claims as most satisfactory a depth of over 100 kms.—from 60 to

70 miles—for his level of complete compensation. All above this consists of the continental rocks. Now it can be easily shown that temperature conditions must arise in such great depths of continental materials as would render them unstable. They would liquefy beneath. The evidence for this will appear in succeeding chapters. There would be only one way out. The continental crust must be composed of rock materials such as never yet have been found at the surface of the earth.

In India and in many parts of Europe, in Africa, the Oceanic Islands, and most recently in Holland, many hundreds of observations have been made, all testifying to the fact of isostatic equilibrium. The evidence shows that a heavy substance underlies the ocean. The earlier attempts at evaluating gravity over the ocean, although supporting isostasy, were not regarded as perfectly reliable. But recently, by observations in a submarine, using a double-pendulum method, invented by Dr. Meinesz, for eliminating the effects of slight oscillations of the vessel, it has been found possible to estimate gravity at sea with much accuracy. Observations carried out in this way in the equatorial regions of the Indian Ocean give results which show that these oceanic areas are in nearly perfect isostatic equilibrium with the continents. As the oceanic layer is of very low specific gravity compared with the continental rocks this result plainly involves an ocean floor possessing a density higher than the continental; as well as the cardinal fact that the continents are, by this same dense medium, supported in a state of flotation.

The mean value of the anomalies for 219 stations over the United States is +0·005 dyne. For 73 stations in India it has been found to be about the same as that obtained for the United States. For 42 stations in Canada the mean anomaly is −0·009. Recent results from Holland closely agree with those from the United States. In various parts of Europe confirmatory observations have been made. It is believed that to a considerable extent the cause of the anomalies is local. The observations in the Indian Ocean as referred to above gave a mean anomaly of −0·011.

In general over the earth the compensation has been found to be so perfect as to have led nearly all authorities to accept the theory of Isostasy, and to claim for it that it is 'fact' rather than theory. Barrell, after a long and critical examination of the results of Hayford and Bowie and of others, considers that: 'The subject remains problematic only in regard to closeness of adjustment and limits of area involved,' i. e. how far the compensation is local or how far it is regional. He states his belief that 'The evidence demonstrates beyond controversy' the essential truth of isostasy.

## APPENDIX TO CHAPTER II

### *Recent Results*

In a recent publication, 'Untersuchungen über Schwerkraft und Isostasie' (Finnisches Geodötische Institut, Helsinki, 1924), Dr. Heiskanen investigates isostatic anomalies not only as observed in Europe and the Caucasus, but also as determined by Hayford and Bowie in the United States. The calculations are made first according to the assumptions of Hayford, and secondly on the hypothesis of Airy as referred to in the foregoing chapter. It is found that the Airy hypothesis is more in accord with the gravity determinations than that of Hayford.

In an article in *Nature* (April 4, 1925), Dr. J. W. Evans discusses this and other recent work leading to the same conclusion and expresses his own views as being in full agreement.

## III

## THE CONTINENTS AND THE SUBSTRATUM

We have every reason to believe that the continents are, in structure, much less homogeneous than the substratum which supports them. The latter, where brought to our observation, by the plateau basalts, is strikingly uniform not only in chemical composition [1] but also in physical structure.

On the other hand, we have abundant evidence at the surface of the earth as to the heterogeneity which prevails beneath our feet. Neglecting the relatively shallow layer of sedimentary materials, we find rocks of deep-seated origin of varying physical and mineral structure. These rocks, while generally agreeing in possessing a highly siliceous composition, present to our view many varieties of structure and mineral composition ; e. g. in the granites, gneisses, anorthosites, syenites, diorites, &c., &c., the granitic type largely preponderating.

A mean density of 2·67 for the continental rocks, based upon their surface representatives, was arrived at by Barrell and is used by Hayford in his calculations referred to in the last chapter. If granites prevail in the depths even more than at the surface—as seems probable—this estimate appears justified : the average density of granites being 2·66.

We must now seek for such information respecting the thickness of the continental crust as may be available. Accurate knowledge is not to be expected; but even approximate estimates are of value. There are certain sources from which information may be derived. Fortu-

[1] See Appendix to Chap. I

nately they involve different data. The fact that they concur in their results is therefore the more reassuring.

We shall first consider the method based on seismological evidence.

The records which reach the seismic observatory reveal two principal types of wave motion as proceeding from the seismic focus or earthquake origin. The cause of the earthquake shock is not certainly known. But, whatever it may be, energy is radiated from it in all directions in the form of wave motions. The fastest moving waves are compressional in their nature—like sound waves. They arrive at the distant observatory first; appearing as waves of small amplitude, and are known as—'preliminary tremors'. These persist till, after a certain interval (which depends upon the distance of the seismic focus), they become merged in waves of considerably larger amplitude.

It is certain that these slower-travelling vibrations represent waves of a different sort. There is reason to believe they are distortional in character; that is, waves which arise from the distortion of the elastic medium and its recoil: such waves as might be generated in a stretched wire by giving it an instantaneous sharp twist.

It has been found that if the seismic focus is distant from the observatory these two classes of waves are clearly separated; the compressional waves arriving first. But distinct separation is only observed at distances greater than about 10° from the origin. Now the separation of the two kinds of waves cannot occur in the heterogeneous media of the continental rocks; for in such media they are continually being reflected and refracted as they pass through rocks of different densities and different elastic properties, and are thereby rendered complex. The separation can occur only in a homogeneous medium, wherein the differing rates of propagation assert themselves.

Waves reaching an observatory 10° distant from the origin will have traversed a considerable distance in the substratum, as shown by the theoretic paths of seismic waves determined by Knott.[1] From this an estimate of 30 km. (a little less than 20 miles) for the thickness of the continents has been deduced by Oldham. Upon similar evidence Wiechert traced continental materials to a depth of 35 km.

A distinctly different method of arriving at an estimate of the mean thickness of the continents depends upon two factors only: the radioactivity of the continental rocks and the amount of heat escaping at the surface as determined by the gradient of temperature. Details will be given in the succeeding chapter. The result arrived at is in accord with the seismic evidence.

Lastly, we saw in the preceding chapter how a continental crust of some 30 km. in thickness floating on a universal substratum of basalt would harmonize with the isostatic equilibrium of oceans and continents.

Thus, we see that the evidence respecting the depth of the continents points to values between 30 and 35 km. Values somewhat lower and somewhat higher have been advocated by writers on this subject.

We have already more than once referred to the view that the continents float on a universal substratum of viscous-solid rock—basaltic in composition—which would exist as a surface rock over about 5/7ths of the globe if the waters of the ocean were removed.[2] We

[1] *The Physics of Earthquake Phenomena*, 1908, chap. xii.

[2] This statement may appear to ignore unduly the oozes and clays which have collected during geological time upon the ocean floor. But

have seen in Chapter II that very strong evidence supports this view. Here we may refer to yet further testimony on this important point.

The velocity of primary earthquake waves across the Pacific floor is found to be appreciably greater than through the continental rocks. The velocities are respectively 3·891±0·028 km. per second and 3·801±0·029 km. per second.[1] Angenheister differentiating between two sorts of surface seismic waves finds that the velocity of the principal waves is 21 to 26 per cent. greater beneath the Pacific than through the Asiatic continent.[2] Wegener, commenting upon these results, claims that they agree with the expected theoretical values obtained from the physical properties of volcanic igneous rocks, and remarks that these differences point to a heavier material composing the ocean floor and—

'It is important to note that we are dealing here essentially with surface waves so that these data become positive proof of the complete absence in the ocean floor of the lighter outermost crust of rock.' [3]

Further indirect evidence arising out of the distribution of terrestrial magnetism is cited by Wegener.[4] This distribution is best accounted for by supposing that the ocean floor is more magnetic than the continents ; a result which would arise if it were composed of basalt.

In short, we possess a consensus of testimony to the existence of a basaltic ocean floor and consequentially of a basaltic substratum. We have the testimony of the

it is not hard to demonstrate from the facts of solvent denudation that the average thickness of these precipitates cannot amount to more than a couple of hundred metres. Joly, *Radioactivity and Geology*, p. 56.

[1] E. Tams, *Centralbl. f. Min. u. Pal.*, 1921.

[2] *Nach. d. Kgl. Ges. d. Wiss. Göttingen*, 1921.

[3] *The Origin of Continents and Oceans*, p. 35.

[4] loc. cit.

great out-flows of basalts during geological time ; of the chemical identity of these basalts ; of isostasy measured over the oceans ; of the transmission of seismic waves beneath the continents and across the ocean floor ; of harmony with the phenomena of terrestrial magnetism. Finally, we have the comprehensive testimony of terrestrial surface history, as will appear when our subject is further developed.

Our next consideration must be the relations of the floating crust with the viscous-solid substratum. In the last chapter we arrived at an approximate estimate of the mean submergence of the continents in the substratum. We found this to be 26 km. (about 16 miles). Upon what is this based ?

We assume that the continents are essentially granitic in composition and the substratum basaltic. We accept for the average density of continental rocks 2·67. For the substratum we accept 3·0, which is the density of gabbro, the deep-seated representative of basalt. We further assume that the pressures and temperatures prevailing will affect both these solid materials in approximately the same manner ; that is, as regards their densities in the solid state. This assumption—i. e. that the ratio of the densities which we observe at the surface is preserved in the depths so long as both media are solid—is, probably, no more than a fair approximation to the facts.

The temperature at the base of the continents, prevailing in the substratum and in the continental rocks, we take to be that of the melting-point of the substratum at that level.

Upon these assumptions we possess in the geodetic measurements of continental elevation a means of calculating the average depth of immersion of the continents in the substratum. We find that the emergent part of the continents (corrected for the buoyancy of the oceans) amounts to 3·22 km. Now the coefficient of buoyancy (as we may call it) arising out of the ratio of the densities of continents and substratum is nearly 8. That is to say, the submerged volume is 8 times the volume of the emergent part. Thus, the submerged depth, or draught, of the continents is about 26 km. If we add the average height of the continents above the ocean floor (which is the surface of the substratum), i. e. 4·62 km., we find the total average thickness of the continental layer to be about 31 km.

As shown in previous pages, we possess *independent* evidence that this may be not far wrong.

We may now, if we so please, accept this evidence as a basis, and regard our calculations as supporting our assumptions as to the densities prevailing in the depths, i. e. the conditions of buoyancy.

Consider what these conditions lead to. A plateau so extended in area as the Tibetan Plateau must be independently compensated. Its average height over the sea is given as 4·575 kilometres (15,000 ft.). Its height above the average continental surface is therefore 4·575–0·820[1] = 3·755 km. Using the coefficient 8, the requisite compensation is nearly 30 km. Even if our coefficient is excessive and we accept one based on a lower estimate of continental density, it can be shown [2] that the temperature within its great compensation due to its own radioactivity must locally attain to 1,500°, which (under surface conditions) would suffice to melt or soften the felspars of the granites; leaving, however, the ground mass of the rock (quartz) rigid.[3] This, however, is an

[1] 0·820 km. is the elevation of the mean continental level over sea level.
[2] *Phil. Mag.*, June 1923.
[3] The melting-point of quartz approximates to 1700°.

extreme case. A plateau much higher than the Tibetan Plateau would be thermally unstable.

Such considerations unfold to us a lurid picture of the dimensions and conditions prevailing in the underworld. It is no dream, however, but arises logically out of prosaic facts, and deductions based thereon.

The continents are very unequal in mean elevation. Europe is estimated to possess a mean elevation above the sea of about 300 metres; Asia 950 m.; Africa 650 m.; Australia 350 m.; North America 700 m.; South America 580 m.; Antarctica 2,000 m. The average oversea elevation of the entire land area is about 800 m. It follows, as their compensations extend into the substratum to depths proportional to those figures, that the Asiatic mean compensation reaches to a depth more than three times that of Europe, and the North American mean compensation is more than twice as deep as the European.

In connexion with the basaltic character of the substratum the physical characters of basalt become of much interest and importance. Unfortunately such data as we possess do not involve pressure conditions such as we are concerned with when we discuss events taking place in the substratum. First, we shall refer to its melting-point.

The experiments of Day, Sosman, and Hostetter, already referred to, show that normal olivine-diabase (a completely crystallized basalt, i.e. free from glass) begins to melt at about 1,150° C. and flows freely at 1,225° C. In the opposite direction—that is when cooling—the melted rock remains fluid down to 1,050°. It then crystallizes suddenly with a sharp increase in density.

In experiments made by the author, using a quite

different method from that used by Day, basalt was found to flow freely at 1,160°–1,170°.[1]

The melting-point will rise with increased pressure, but not indefinitely.[2] By analogy with the behaviour of certain organic substances, it may be inferred that the melting temperature, after increasing with the pressure up to a certain point, will ultimately decrease with further augmentation of pressure. But, for the silicates, the pressures required to bring about this change must be very great. This stage is only reached when the volume change attending fusion is no longer positive ; that is when the pressure (acting more effectively on the liquid than on the solid) is sufficient to control an increase of volume. Vogt concludes that the pressure due to 25 miles of rock might raise the melting-point by 50° ; and that the maximum melting-point might be reached at a depth of 150 km. (94 miles).[3]

One of the most important applications of our knowledge (such as it is) of the physical conditions affecting the flotation of the continents refers to the change of buoyancy of the magma which must arise if the substratum changes from the solid to the liquid state. For it will, in fact, appear that although it is at the present time in the solid state, yet in the past it has been liquid, and must inevitably revert to the fluid state in the remote future. How will this change affect the 'freeboard' of the continents ?

When we melt basalt in the laboratory we easily float granite in the fluid rock. It floats with a good freeboard, and appears to preserve its buoyancy indefinitely.

[1] 'On the Volume-change of Rocks and Minerals attending Fusion,' *Trans. R. Dub. Soc.*, vol. vi, 1897.

[2] See Appendix to this chapter.

[3] Vogt, loc. cit., p. 208. See also Harker, *Natural History of Igneous Rocks*, pp. 163–4.

The feldspar is found to be partially fused, but the quartz which acts as ground-mass is unaffected.

Referring again to the results obtained by Day, Sosman, and Hostetter, we find that the difference in volume between crystalline and liquid basalt at 1,150° (at which temperature the crystalline rock begins to fuse) is, as observed in the laboratory, about 12 per cent. of the volume of the crystalline rock at that temperature. Lesser results have been obtained by Barus (see Appendix to this chapter), and by the author, who used a different method of investigation.

A rough estimate of the effects of a volume-change of 12 per cent. suffices to show that—as would be anticipated—the attendant loss of density is considerably more than adequate to explain the amount of sinking of the continents as inferred from geological history.

Now it is well known that liquids are much more compressible than solids. Liquids are an intermediate state between the solid and the gaseous. Their compressibility is of a different order from that of solid compressibility. Thus the compressibility of solids—the reduction of volume per megabar[1]—may be from $0{\cdot}5 \times 10^{-6}$ (steel) to $3{\cdot}0 \times 10^{-6}$ (glass); whereas that of liquids may be $78 \times 10^{-6}$ (turpentine); $62 \times 10^{-6}$ (olive oil); $25 \times 10^{-6}$ (glycerine); $43 \times 10^{-6}$ (water).

We gather from this that we cannot ignore the effects of pressure on the liquid basalt, although it is allowable to do so when considering the volume ratio of solid granite and solid basalt.

As the compressibilities of most liquids are round about 0·00005 per atmosphere (which, if sustained, would amount to 5 per cent. for 1,000 atmospheres), we may

[1] The megabar is one million dynes per sq. cm. It is nearly one atmosphere.

reasonably expect that the density of fluid basalt will, at a pressure of over 8,000 atmospheres, be increased by at least a few units per cent.

Taking this into account, it is possible on certain permissible assumptions to arrive at an idea of the amount of continental flooding which might arise. From our previous estimate we found that the continents are at the present time submerged to a depth of 26 km. in the substratum, now solid and having the density 3·00. The emergent part of the continents measured to the mean continental surface is 4·62 km. The mean sea depth is 3·80 km. (Murray), and the elevation of the mean continental level over sea level is 0·82 km.

Now when the substratum melts its density falls. We shall assume the density-loss to be 7 per cent., i. e. the density becomes 2·79. The effect on the elevation of the continents will be gradual, because at first when melting begins the ocean floor is thick and encloses the continents so that they do not feel the loss of buoyancy; but as the floor gets thinner the continents feel more and more the effects of the density change. We shall suppose the floor to diminish to a minimum thickness of about 6 km., which will leave about 20 km. of the continents immersed in the melted substratum. As the density of this is 7 per cent. less than that of the viscous-solid basalt previously buoying up the continents, these must sink proportionately. That is, they sink 7 per cent. of 20 km. = 1·4 km. The mean surface of the continents would be under water to a depth of 1·4—0·82 = 0·58 km. As the sea flows in upon the continents its surface sinks, and its surface also sinks owing to the increased area of the ocean floor prevailing at this time.[1] These corrections are small, and are offset by the effect of the weight of the

[1] See Chap. VI.

transgressional seas. The total correction may be taken as leaving a downward movement of 1·4 km. (4,600 ft.) relative to the prevailing ocean surface.

According to the geodetic data of Wagner,[1] this would involve a loss of continental area of about 80 per cent. This, although it is a mere approximation, suggests that the effective loss of density of the substratum is probably less than the assumed 7 per cent. The depths and areas of transgressional seas are not accurately known. Vertical movements of large continental areas considerably greater than 1·4 km. are on record as having occurred in late Tertiary and early Pleistocene times.[2] The palaeogeographic maps of Willis[3] of late Middle and Upper Cambrian, Middle Ordovician, Silurian, and Middle Devonian show continental seas covering by far the greater part of North America. Grabau's maps[4] show even greater flooding in Palaeozoic times, and almost complete submergence of North America in Cambro-Ordovician times. The series of palaeogeographic maps prepared by Arldt from the investigations of many authorities support these views and indicate that in Eurasia very similar conditions obtained in Ordovician and Silurian times.[5] Some authorities, however, hold the view that the loss of more than 50 per cent. of the land area at the one time is improbable.

We may now again recall the statement which will be found on p. 29 that the continents are for the greater part covered with materials which plainly indicate that in past times they must have been largely under the surface of the ocean. From what has been said above it will be evident

[1] Supan, *Grundzüge d. phys. Erdkunde*, p. 47.

[2] Willis and Salisbury, *Outlines of Geologic History*, Chicago Press, 1910, p. 265 et seq.

[3] loc. cit., p. 41 et seq.

[4] loc. cit., p. 44 et seq.

[5] Arldt, *Handbuch der Palaeogeographie*, 1919.

that if the substratum were to melt and if the volume-change attending the change of state were even as little as 4 or 5 per cent., there would be a downward movement which would be sufficient to account for the continental flooding. Later we shall further consider this explanation.

There is conclusive evidence that the substratum at the present time is solid.[1] Towards rapidly changing stresses it behaves as a rigid, elastic substance. Seismic waves reaching an observatory from distant earthquake shocks afford unequivocal evidence of its solidity. For, as before explained, these waves are to a large extent distortional in character, and such are not transmitted by fluid media. However, it is certain that after passing through many hundreds or thousands of miles, and descending to considerable depths in the substratum, they preserve their identity and conserve their energy. The conclusion is that no continuous layers of fluid matter can be mingled with the substratum.

But there is further evidence for the solidity of the substratum : evidence derived from the tidal effects due to solar and lunar gravitational attractions.

In tidal theory it is shown that a fluid envelope covering the earth must be subjected to bodily displacement due to the attraction of sun and moon. It is sufficient to consider the more effective of these forces—that due to the moon.

If we suppose a fluid layer, composed of a frictionless liquid, to cover the earth, the tidal protuberances would maintain positions beneath and opposite to the moon, as

[1] It would seem, however, to be viscous towards great and long-continued forces. Hence it probably responds to such forces, and so isostatic equilibrium prevails with respect to the larger surface features of the earth.

in Fig. 5; the fluid medium everywhere flowing into the tidal wave, which would confer upon the earth an egg-shaped form, with the ends of the egg pointing towards and away from the moon. If, however, the fluid is not frictionless, then the tidal wave is delayed in its genesis and it is carried by the rotation of the earth into the positions shown in Fig. 6.

We now notice that the attraction of the moon on the tidal protuberances must affect the rotation of the earth; tending to retard its axial velocity. For although the pull on the tidal protuberance remote from the moon acts

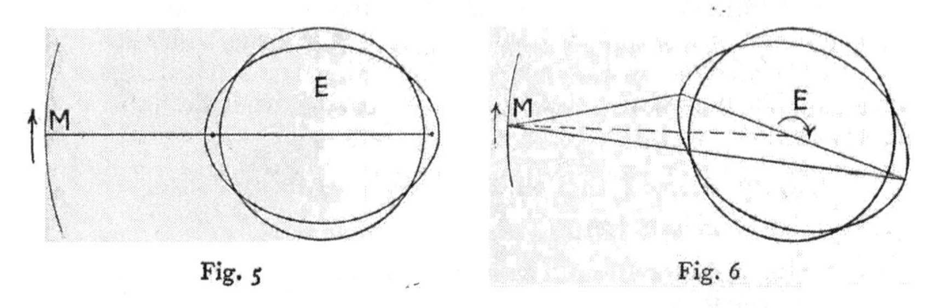


Fig. 5    Fig. 6

the other way, the most effective pull is that applied to the protuberance on the side nearest to the moon.

Fig. 6 applies to tides generated in a fluid layer of very great depth. In the case of the oceanic tides not high water but low water would be under and opposite to the moon. But in this case also the tidal friction will tend to retard the axial rotation of the earth. The reason for this is that frictional effects in the case of a shallow layer, such as the ocean, cause the tidal protuberance to be displaced backward, i. e. west to east. The retardative effect therefore acts in the same direction as in the case of a deeper fluid layer. The effect is as if a brake were applied to the rotating earth. Its period of rotation must slowly augment and the day grow longer. Now there is evidence that since remote historical times there can have been only very little lengthening of the day, so little as to be detected only with difficulty from the records of ancient eclipses. It is lengthening about one-thousandth of a second per century or one minute in six million years.

Love, in a recent communication to the Royal Society, discusses the matter in the light of recent additions to our knowledge and concludes :

'It appears, however, to be quite possible that the earth may consist of a very dense and very rigid core enclosed in, and connected by solid matter with, a lighter shallow crust, the greater part of which is solid and of a rigidity comparable with that of granite (about one-third of that of steel), the shell being honeycombed with hollow spaces containing molten matter. But it seems to be impossible that the molten matter should form a continuous layer separating the outer portions of the earth's body from the inner features.' [1]

Darwin closes the consideration of the question in equally decisive terms :

'Although there are still some—principally among the geologists—who believe in the existence of liquid matter immediately under the solid crust of the earth, yet the arguments which I have sketched appear to most men of science conclusive against such belief.' [2]

The tides, from another line of research, afford further evidence upon the matter. The results of careful systematic measurements applied to the oceanic tides have been shown by Kelvin and Darwin to support the view that the earth is rigid as a whole. For it has been mathematically deduced that if a fluid substratum existed, or the earth were not highly rigid, the oceanic tides would not possess their present amplitude. The explanation is that if the earth itself became tidally deformed the oceanic tides must become purely differential in magni-

[1] *Proc. R. S.*, A, vol. 82 ; *Nature*, April 29, 1909.

[2] The reader is referred to Darwin's *Tides*, and to an Address by Eddington, *Nature*, January 6, 1923.

tude and, in fact, would exist merely as a residual effect. In a certain degree they are, as it is, of this character in consequence of the body of the earth not being perfectly rigid. They would lose in definition very much more if the substratum of the continents were fluid.

That the substratum is solid at the present time may be accepted as certain. Later it will appear that it is equally certain that it is highly heated.

Knott, dealing with the subject of 'Earthquake waves and the Elasticity of the Earth',[1] says with reference to the apparently contradictory evidence of volcanicity and its fluid ejecta and the types of wave-motion known to be transmitted through the substratum:

> 'Whatever be the nature of the materials lying immediately below the accessible crust, it must become at a certain depth a highly heated fairly homogeneous substance behaving like an elastic solid, with two kinds of elasticity giving rise to what are called the compressional and distortional waves.'

Have we any knowledge as to the depth to which the substratum extends? The answer is that seismology has here again come to our aid and, although it has not fully answered the question, has given us valuable information.

In his anniversary address to the Geological Society in 1922, Oldham expresses his belief 'that the real and ultimate origin of earthquakes must be more deeply seated than the surface rocks'. And later, in the case of an earthquake in Northern Italy, he traces the origin to a depth of the order of 100 miles below the surface.[2] Turner also holds the view that the origin of many earthquakes giving good long-distance records lies at a depth of about 125 miles, and cites cases at depths varying from

[1] *Abstr. Proc. Geol. Soc. of London*, June 15, 1920, p. 84.
[2] *Q.J.G.S.* lxxix, 1923.

50 to 300 miles below the surface. Pilgrim estimated the origin of an earthquake at 100 miles beneath the surface. Banerji questions Turner's estimate of depth and concludes that the maximum depth derived from recorded results must be less than 200 km.; but may be anything in the neighbourhood of 100 km.[1]

There can be little doubt that these origins lie in the substratum. But whether the basaltic region is carried so far down is, of course, not safely inferred. We know of no silicated rocks denser than the peridotites and allied magmas, which, as we have already seen, are of the nature of those ultra-basic silicates found in stony meteorites. A metallic terrestrial core must lie deeper. It seems probable that the basaltic shell persists some one or two hundred miles beneath the base of the continents, and that an ultra-basic shell succeeds it. But as we go deeper our views become increasingly speculative.

In times to come we may know more about the substratum. There is every incentive to investigation for the changes progressing within it have controlled the surface history of the earth.

[1] *Phil. Mag.*, Jan. 1925.

## APPENDIX TO CHAPTER III

### *The Effects of Pressure upon the Melting-point*

A well-known thermodynamic equation due to James Thomson exhibits the relation between melting-point and pressure:

$$\frac{d\theta}{dp} = \frac{\theta\,(V_2 - V_1)}{L}$$

where $\theta$ is the melting-point reckoned from absolute zero (i. e. $t$°C. + 273°) and $p$ is the pressure in atmospheres (1 atm. = 1,033 grm. per sq. cm.). $L$ is latent heat of fusion in calories expressed as work (1 grm. cal. = 42,800 grm. cm.). $V_2 - V_1$ is the increment of specific volume (i. e. the increase in volume which one grm. experiences on melting, expressed in cub. cm.).

This equation leads to certain valuable general conclusions, but cannot be applied to great depths because the volume-change and the latent heat vary with the depth in a manner not as yet ascertained.

In its results we find that a positive volume-change attending fusion (i. e. an expansion) involves a rise of melting-point attending increase of pressure: a negative volume-change (i. e. a contraction) involves lowered melting-point attending increase of pressure. Thus the melting-point of ice, which is one of the few substances with a negative volume-change, is lowered 0°·0075 C. for an increase of pressure of 1 atmosphere. In the great depths of the ocean the melting-point must be some 7 or 8 degrees lower than at the surface. That is to say, fresh water would not freeze at these depths till it was cooled to 7° or 8° below 0° C.

The rocks exhibit a positive change of volume when fused. Hence they would sink in their own melts. Under very high pressures this would not be expected to happen, for the compressibility of liquids is much greater than that of solids. Hence such substances as rocks would be expected to possess a maximum melting-point; i. e. that corresponding to a pressure which renders the volume-change null. The experiments of Tammann on certain organic substances revealed this condition; i. e. $\frac{d\theta}{dp}$ became equal to 0. Doelter thinks that the maximum melting-point of a rock melting normally at 1,100° might lie between 1,300° and 2,000° and might be reached at depths of from 150 to 370 km. (90 to 230 miles).

It is not profitable to speculate upon what may happen at still greater depths. But it may be kept in view that conditions such as would enable thermal energy to be stored without limit are inconceivable. Instability must ultimately set in. Release of energy necessarily involves upward movement. This must sooner or later be brought about and loss of rigidity of the overlying parts of the substratum, progressing downwards from above, must be favourable to such movements. Hence the deep-lying energy of the substratum would be expected to find relief in translational movements upwards at intervals in geological history, as will be shown in later chapters.

(See Doelter, *Handb. der Min. Chem.* B. i, 1912, p. 674, and Harker, *Nat. Hist. of Igneous Rocks*, Methuen, 1909, p. 163.)

### *The Change of Quartz into Cristobalite and Tridymite*

There is one respect in which our assumption as to the density of the continental rocks where exposed to high temperature may err. It is known that quartz which makes up from 25 to 35 per cent. of the granites[1] undergoes an allotropic modification at 1,000°, at atmospheric pressure. It changes into cristobalite—which is a low-density form of quartz. In presence of water it may, at 870°, change to tridymite—also a substance of lower density than quartz.[2] Day, Sosman, and Hostetter have measured the volume-change of quartz over a range of temperature extending to 1,600°. The fall in density at high temperature is conspicuous. They also investigated the volume-change of granite, and here also the effect of the change of the quartz to cristobalite is recognizable. At 20° the specific volume of cristobalite is 13·4 per cent. greater than that of quartz. If, then, the quartz in a given granite were changed to cristobalite, the volume-increase, or fall in density, should be some 4 per cent. from this cause only.[3]

If this change obtains in the depths where the continental rocks are at a temperature over 1,000°, the density at these lower levels would be not 2·67 but about 2·56. But we cannot without further evidence accept this density. It may be that this change will not occur under the great pressures—some 8,000 atmospheres—prevailing at these depths. And the fact that cristobalite—which is a stable substance—is not found in the granites which reach the surface, nor in any similar rocks, supports this view.

[1] Sollas, *Trans. R.I.A.* xxi, 1891.
[2] Fenner, *Am. J. Sc.* xxxvi. 331, 1913.
[3] *Am. J. Sc.* xxxvii, 1914.

### *The Sinking of the Continents attending Liquefaction of the Substratum*

The calculations given in Chapter III respecting the sinking of the continents may be more directly referred to experimental data by utilizing the results of Douglas and Barus. Daly has rendered these available in a convenient form by calculating the specific gravities of certain important rocks at different temperatures, both when in the solid and the molten state.[1]

Referring to these we find that gabbro, of which the density at 20° is 3·00, has the density 2·92 at 1,000°; while granite, of the density 2·70 at 20°, has the density 2·63 at 1,000°. This would be, closely, 2·60 for a rock of density 2·67 at 20°, which affords the coefficient of buoyancy 8, as assumed in our calculations. Variants around this figure may be found by selecting rocks differing in initial density. From this we find that so far as experimental measurements can help us there is no reason to discard this coefficient.

When we apply Daly's tables to the problems arising out of change of state of the substratum, we find that gabbro, which has the density 3·0 at 20° and the density 2·92 at 1,000°, assumes the density when molten of 2·74 at 1,100° or 2·75 at 1,000°. The percentage-loss of density is 6·0 very nearly. And when we further examine the tables we find this percentage fall in density is in many cases approximately arrived at.

It is of interest now to estimate the amount of continental flooding arising out of this figure.

The buoyancy of the magma falls 6 per cent. If that part of the submerged continents affected is 20 km. deep, then the continents sink 1·2 km. The emergence above the substratum becomes 3·42 km. and, as the sea-depth is 3·80 km., the mean level is submerged 0·38 km., i. e. 380 metres. The actual sinking of the coast-line will be about 1·2 km., and from the tables of Wagner as quoted in Chapter III we find that about 75 per cent. of the continental area is submerged.

This calculation takes no account of the effects of pressure which *must* diminish the loss of density attending change of state. *It is evident that on the data at our disposal we are justified in ascribing the epeiric seas, attending the coming of a revolution, to the density-changes arising from the change of state of a basaltic substratum.* Further than this we cannot go, at present.

[1] *Igneous Rocks*, p. 202.

## IV

## THE RADIOACTIVITY OF THE ROCKS

We have now to enter upon a subject which is quite as fundamentally important in the surface history of the earth as isostasy itself.

The rocks of the earth's surface, the granites, the gneisses, &c., composing the continents, and the plateau basalts rising from the substratum, are one and all radioactive. That is, they contain minute amounts of those elements which are continually and automatically changing into forms of lower atomic weight, and developing heat in the act of transformation. The lost mass, expelled in atomic form, is radiated with great violence; but is soon brought to rest by surrounding atoms. In its arrestment heat is developed, just as the target is heated in arresting the flight of the bullet. Simultaneously the recoil of the parent atom is also arrested. In this way, principally, the phenomenon of radioactivity becomes a slow but enduring form of thermal genesis in the rocks.

The radioactive substances are all traceable to two parent elements—the very heavy elements uranium and thorium. These are the heaviest elements known. It is not improbable that their instability in some way arises out of their great atomic weight. They very slowly break down; gradually losing mass, stage by stage, till they assume the properties of lead.

In the course of disintegration the atomic weight of uranium descends from 238 to that of lead having the atomic weight 206; and in the case of thorium the loss of atomic weight is from 232 to lead having the atomic weight 208. Ordinary lead, which may possibly be a mixture of uranium-derived and thorium-derived lead, has an atomic weight of 207.

We have to recognize that radioactive phenomena are entirely different from chemical ones. The former are concerned with spontaneous internal changes progressing within the atom and probably originating in the atomic nucleus. The material atoms which are radiated during atomic disintegration are an integral portion of the parent atom. The alpha ray is in fact a positively electrified helium atom derived from the nucleus. Chemical actions, on the other hand, are concerned with the inter-relations of molecules. Hence they are readily affected by such forces as may affect the concentration of the material particles; their state of thermal agitation; and above all their ionization.

But it is quite otherwise with radioactive events. Neither heat nor pressure seems to affect them in the least. Fajans writes:

> 'Attempts have been made to influence the rate of transformation of emanation and many other radio-elements, but even increasing the temperature from −250° to over 1,000° C. has not the slightest effect on the velocity of transformation or the intensity of the radiation of a radio-element. Again, the most intense treatment with X- or cathode-rays or with the rays of radioactive substances themselves is of no avail. So to-day we are still powerless against the natural radioactive transformations; we can neither accelerate nor retard them.' [1]

Bronson [2] used extremes of temperature from −180° C. to 1,600° C. without effect. H. W. Schmidt [3] found that neither temperatures of 1,500° C. nor pressures of 2,000 atmospheres produced any change in the radioactivity of radium preparations. Rutherford and Petavel tried if the pressure and temperature arising from an explosion of cordite (about 2,500° C. and 1,200

[1] *Radioactivity*, p. 13. [2] *Proc. R. S.* 78A.
[3] *Phys. Zeit.*, February 15, 1908.

atmospheres) would affect the radiations passing through the steel walls of the explosion chamber. During the explosion there was no effect. A small temporary effect of 9 per cent. in the radiation was indicated after the explosion; possibly arising from stress-effects in the vessel.

Although uranium and thorium are scarce elements in the rocks it will take a very long time before they are even one-half used up. If our estimates of their rate of decay may be applied over the future, one-half of the uranium now upon the earth will have disappeared in about 5,000 million years; and one-half the thorium will have disappeared in about 13,000 million years. Now the lapse of time since the oldest sediments were laid down in the primeval oceans may be some 200 millions of years or as some think 800 millions of years or even longer. Hence even if we accept extreme estimates of the age of Archaean time (1,200 millions of years), geological time is but a small fraction of the period it will take to reduce sensibly the influence of radioactivity.

The quantities of uranium and thorium in the rocks are very minute, so much so that ordinary chemical methods would fail to estimate them. But by taking advantage of the special physical properties exhibited by certain of the elements into which they change, it is possible to deduce with considerable accuracy the amount of the parent elements.

As the result of some hundreds of measurements it is known that all rocks found on the surface of the earth contain these parent substances. The granites and similar acidic rocks are the richest in uranium and thorium. The more basic rocks—the basalts—contain considerably less. The secondary or sedimentary rocks are referred to in the Appendix, p. 77.

These measurements deal with such minute quantities

that they require to be stated in billionths of a gramme : that is in millionths of millionths : conveniently written $10^{-12}$ grm. It is convenient to estimate the quantity of uranium indirectly : the quantity of one of the derived elements, radium, being obtained in the first instance from the electrical properties of a gas (the 'emanation') into which radium is transformed. Thus the experiment gives us in the case of granites an amount which in the mean will be $3 \times 10^{-12}$ grm. of radium per grm. of granite. It is now easy to find from this the amount of uranium present, if we so desire, for there is an invariable relation between the two quantities.

In the case of thorium it is customary to express the amount of thorium directly, although it also is measured not directly as thorium but in terms of a derived element, also a gas—known as thorium emanation. Thus a granite might be expected to contain $2 \cdot 0 \times 10^{-5}$ grm. of thorium per grm. of rock. Similarly the average radium content of a large number (58) of basalts from all parts of the earth recently examined was found to be $1 \cdot 19 \times 10^{-12}$ grm. per grm. and the average thorium content was $0 \cdot 77 \times 10^{-5}$ grm. per grm.[1] These are important figures. But perhaps they are not so important as the results arrived at for the great basalt outflows. Specimens from three great outflows—the Deccan, the Hebridean, and the Oregonian—were examined. The following table contains these results :

| | Radium | Thorium |
|---|---|---|
| Deccan (6) | $0 \cdot 77 \times 10^{-12}$ | $0.46 \times 10^{-5}$ |
| Hebridean (6) | $0 \cdot 77 \times 10^{-12}$ | $0 \cdot 49 \times 10^{-5}$ |
| Oregonian [2] (7) | $1 \cdot 69 \times 10^{-12}$ | $1 \cdot 52 \times 10^{-5}$ |

It is easy to calculate from these figures the quantity of heat being continually developed in any of these rocks.

[1] J. H. J. Poole and J. Joly, *Phil. Mag.*, November 1924.

[2] The authenticity of this material as part of the true plateau basalt is not perfectly assured. *Phil. Mag.*, loc. cit.

Thus it has been determined by special investigations that the heat continually evolved by one gram of radium 'in equilibrium' with all its related elements (i. e. the whole series from uranium downwards) is $5 \cdot 6 \times 10^{-2}$ calorie per second.[1] Also the heat evolved by one gram of thorium and related elements is $6 \cdot 6 \times 10^{-9}$ calorie per second.

In the case of granite, therefore, one gram of the rock develops $3 \times 10^{-12} \times 5 \cdot 6 \times 10^{-2} = 16 \cdot 8 \times 10^{-14}$ calorie per second from the uranium series of elements present; and $2 \cdot 0 \times 10^{-5} \times 6 \cdot 6 \times 10^{-9} = 13 \cdot 2 \times 10^{-14}$ calorie per second from the thorium series of elements present. Adding these quantities we find that the heat developed per second per grm. of rock is $30 \cdot 0 \times 10^{-14}$ calorie. This is an extremely small quantity but then a second is a short time, and there are a great many of them in a million years, which is our unit of time when we deal with geological events.

Now if we reflect upon this, we must conclude that here is something which must certainly have affected the surface history of the earth. For the outflow of heat is unceasing throughout the unrealizable vistas of geological time.

The present Lord Rayleigh first rendered quantitative considerations possible by his measurements published in two papers in 1906.[2] Quantities of radium of the order cited above were definitely determined in a large number of rocks, both igneous and sedimentary. In these papers the very natural suggestion was made that such quantities of heat-producing substances could not extend deeper than a very few miles from the surface of the earth. It must be

[1] The calorie referred to is the quantity of heat required to raise the temperature of one gram of water by one degree centigrade.

[2] Strutt, *Proc. R. S.* 77A and 78A.

remembered that at this time no systematic measurements of the thorium content of the earth's rocks had been attempted. A method of effecting these measurements was devised and many rock substances were examined in 1909.[1] The argument for a limited extension downwards might appear to be strengthened by the increase to the heat-producing power of radioactive substances in the rocks due to thorium. But an arbitrary and unaccountable limitation of radioactivity is, as we shall see, unnecessary.

Let us now examine, so far as we may be able, the existing thermal state of the earth's crust. It is known that when we bore down into the surface rocks the temperature rises. The rate of rise—the temperature gradient—varies considerably from place to place. The fact of a gradient existing in the continental layer involves the continual escape of heat at the surface. It is easy to estimate the amount of heat which escapes.

We shall assume that the continents possess an average depth of 31 km. (20 miles) and that they are composed of rocks of granitic character possessing the radioactivity estimated above—i. e. giving rise to $30 \times 10^{-14}$ calorie per grm. per second. If their density be 2·7, this works out as about $0{\cdot}8 \times 10^{-12}$ calorie per cubic centimetre per second. If no heat passes down through the floor of the continent into the supporting magma the heat developed throughout the entire depth must be continually escaping at the surface.

Consider a vertical column one square centimetre in cross-section. Obviously the quantity of heat developed per second in this column is $0{\cdot}8 \times 10^{-12} \times 31 \times 10^{5} = 2{\cdot}48 \times 10^{-6}$ calorie. This quantity must be escaping every second at the surface. How does this agree with the observed gradient?

[1] Joly, *Phil. Mag.*, May 1909.

Gradients are very various. They may be said to vary from 1° C. in 28 metres to 1° C. in 40 metres. The gradient often accepted as most representative is 3·5° C. in 100 metres; that is to say, the temperature rises on the average 3·5° C. for each hundred metres descent. This in itself is not an accurate statement; for, according to readings obtained from deep borings, the gradient steepens a little as we go down. This rise, probably, ceases after a certain depth is reached.

The quantity of heat escaping at the surface may be readily calculated if we know the conductivity of the rocks in which it is observed. The conductivity of a substance may be defined as the quantity of heat which a slab of it, cut to the thickness of one centimetre, will transmit per square centimetre in one second if a temperature difference of 1 degree centigrade is maintained on the opposite faces. Thus a slab of granite under these conditions will transmit about 0·005 calorie per square centimetre per second. Some granites transmit rather less. It is common to take 0.004 as the conductivity of the continental rocks in general.[1]

Basalt has a conductivity of 0.004. Moisture seriously affects the conductivity of average porous rocks. Thus sandstone (a common surface rock) when dry has a conductivity of 0·0055; but when damp its conductivity may increase to 0·0085. On the average, damp sandstone has the conductivity 0·007.

The quantity of heat conducted varies with the gradient. Where there is no gradient heat cannot flow. The gradient of 1° across one centimetre is unit gradient. If it is 2° per centimetre twice as much heat will pass

[1] Experiments on granite and basalt exposed to temperatures up to 600° reveal a lowered conductivity due apparently to development of flaws. H. H. Poole, *Phil. Mag.*, Jan. 1914.

through—and so on. In such units the gradient at the earth's surface taken at $3 \cdot 5^{\circ}$ in 100 metres is, evidently, $3 \cdot 5^{\circ} \div 100 \times 100$. We may now calculate the outward heat flow at the earth's surface. We may take it that the rocks in which the gradient has been observed are more or less damp. It is well known that what is called 'quarry sap' exists in every surface rock. We therefore accept 0·007 as the conductivity. We multiply this by the surface gradient: that is by $3 \cdot 5 \times 10^{-4}$. This gives us $2 \cdot 45 \times 10^{-6}$ calorie per second per square centimetre, which agrees with $2 \cdot 48 \times 10^{-6}$ found independently above.

Thus we find that the observed gradient of temperature at the surface of the earth is in harmony with the view that the observed radioactivity of the rocks prevails throughout the whole depth of the continental layer.

We have assumed in these estimates that no heat enters from beneath into the continental crust. We know that the basalt beneath must be radioactive. But if heat is not escaping from it then we must assume that its present temperature is just that prevailing at the base of the continents. We have to anticipate here. Later we shall find reasons for believing that the substratum is now much in the condition of recently solidified molten basalt. Now Day, Sosman, and Hostetter observed that when basalt cooled from a state of fluidity the temperature fell till 1,050° was reached. Then the liquid basalt suddenly crystallized. It is certain that the temperature of crystallization will not be quite the same under the pressure-conditions prevailing beneath the continents. In fact we may conclude that both melting and freezing will take place at somewhat higher temperatures than those observed in the laboratory. But as we are only dealing with approximations we shall adhere to the laboratory results.

Our argument is that if no heat from the substratum is ascending through the continental layer and if it be true —as already assumed—that no heat from the continental layer is escaping downwards into the substratum, then the temperature of the basalt and the temperature at base of the continental layer must be the same. That of the substratum may be at the present time in or about 1,050°. What is the average temperature at the base of the continents as calculated from the radioactivity of the continental rocks?

It is not difficult to arrive at an estimate of the temperature prevailing at the base of the continental layer if no heat is escaping downwards. If $\theta$ be this temperature it can be shown that $\theta = \frac{Q \times D^2}{2K}$, where $Q$ is the quantity of heat of radioactive origin produced per second in a cubic centimetre, $D$ is the thickness of the continental crust in centimetres, and $K$ is the conductivity.[1] Of these data we know $Q$ to be approximately $0{\cdot}8 \times 10^{-12}$ calorie as found above: $D$ we take as $31{\cdot}4 \times 10^5$ cm.: $K$ may be taken as $4 \times 10^{-3}$. We find $\theta°$ to be 986° C.

This result further supports the view that the continental crust transmits heat to the surface which is mainly derived from its own radioactivity. *There are no difficulties in the way of this view provided we can account for the heat continually being generated beneath in the substratum.* Later we shall see what becomes of this heat. Independently of the result arrived at above, we have reason for believing that some of it does in fact enter the continental crust. Where the thickness of the crust is less than the mean thickness, or approximates to the mean thickness, this probably happens. Where the continental thickness is much above the average the opposite must happen. Some part of the heat then flows downwards into the substratum.

[1] Strutt, *loc. cit.*

We also see that our results may be regarded from another point of view. We proceed on an estimated continental depth which is about the mean of the seismic results, and upon a radioactivity of the rocks based on some scores of experiments, and we find a basal temperature for the continental layer which is about that of recently solidified basalt. Is this accidental? Further on reasons will be given which go to show that so far from being accidental it is inevitable.

A further and obvious deduction from our results is that the substratum beneath the continents must now be accumulating its own radioactive heat. For, otherwise, where is it to go? There is no evading this conclusion. It cannot be passing upwards to the terrestrial surface for the continental radioactive heat accounts for almost all that is escaping at that surface. Slowly throughout the passing ages the heat derived from the disintegrating atoms is accumulating in the substratum. What will become of it?

Beneath the oceans the conditions are somewhat different. The substratum comes directly in contact with the cold ocean water. Such radioactive heat as is being generated in its upper parts must be passing into the ocean by conductivity. But this condition cannot prevail indefinitely downwards. At a certain depth the basal temperature of the cooling basalt will have risen, *by its own proper radioactivity*, to that of its melting-point. A layer of basalt possessing a basal temperature from its own radioactive heat the same as that of the underlying basalt (i.e. the melting-point at that depth) must conserve the heat beneath it just as the continental crust conserves the heat of the subcontinental magma. Hence, from the deeper parts of the substratum, no heat can be escaping through the ocean floor; but the heat must be accumulating just as it accumulates beneath the continents and for the same reason. The critical depth beneath which all heat must be conserved will be about 48 km. (30 miles).[1] As will be seen later, it is probable that at the present time heat is passing into the oceans from the upper parts of the solid but highly heated substratum, the ocean floor being still undeveloped. The floor throughout the whole period of its development is at its upper surface at the temperature of the water (0° C.) and at its base at the melting-point of the basalt proper to the depth. Its mean temperature, therefore, varies but little.

We here close our study of the radioactive conditions affecting the floating continents, the ocean floor, and the substratum. It is quite evident that the thermal conditions cannot remain as they are. The radioactive heat is for ever accumulating.

We find, in fact, that we are dwellers upon a world in the surface materials of which there exists an all but inexhaustible source of heat. If the natural question be asked, 'Does this source of heat also extend into the interior?' we must answer that such evidence as we possess suggests an interior resembling the nickeliferous iron of the meteorite. Meteoric iron sometimes contains a minute trace of radium and therefore of uranium. The thorium content of metallic meteorites has not been investigated. In terrestrial rocks there is a certain rough proportionality between the uranium and thorium content. They are *about* as two to one. But this may not apply to meteoric iron. All siliceous meteorites contain radium. Some have a radium content approaching that of similar stony materials on the earth, e. g. the ultra-basic rocks; and, in the one case examined, a very small amount of thorium.[2] It seems probable from

[1] Cotter, *Phil. Mag.*, September 1924.

[2] See a paper by Quirke and Finkelstein, *Am. J. of Sc.*, September 1917; and one by the author, *Phil. Mag.*, July 1909.

these facts that the radioactivity of the magma is succeeded by the lesser radioactivity of an ultra-basic layer which may reach to a depth of several hundred miles from the earth's surface. Farther down, the great core of the earth probably possesses much smaller quantities of the heat-producing elements. But this is surmise.

Looking back into the past we may ask, 'Has there ever been any other source of heat than that of radioactivity? If there has, what is the evidence for it?' Whatever be the answer to these questions we are certainly entitled to give our first consideration to the effects of that source of heat which we know exists. Those who think that a radioactive substratum is speculative have—if the claims of isostasy are to be satisfied—to substitute some other heavy rock instead of basalt, possessing physical properties similar to those of basalt as regards density and melting-point, but possessing no trace of radioactive elements. Such a substance is unknown at the surface of the earth.

## APPENDIX TO CHAPTER IV

### *The Radioactivity of the Secondary Rocks*

The sedimentary rocks (e. g. slates, shales, sandstones, limestones, &c., &c.) may in amount be equivalent to a layer about two kilometres deep spread over the continents. The average radioactivity of these rocks is, as might be expected, less than that of the primary (igneous) rocks from which they are derived. The balance has been carried by solvent denudation into the ocean and is, probably, responsible for the high radioactivity of some of the deep sea deposits.

As the result of a very large number of measurements the average evolution of radioactive heat (from the uranium and thorium series) per grm. of the sedimentary rocks is $16\cdot6 \times 10^{-14}$ calorie; as contrasted with $30 \times 10^{-14}$ calorie in the granitic materials probably making up the greater part of the continents.[1]

[1] See Joly, *Phil. Mag.*, October 1912; and papers by J. H. J. Poole and A. L. Fletcher in the same Journal.

## V

## THE DECIPHERMENT OF SURFACE HISTORY

EXAMINATION of the earth's surface materials reveals on every side what at first sight appear to be irreconcilable facts and hopeless confusion. Many hundreds of feet above sea-level we find rocks which were certainly formed in deep water; others which were deposited in shallow water. Elsewhere such sediments are piled up to form great mountain ranges. Moreover, there is evidence of very great compressive forces in the folding and deformation of these sediments, and we know that these forces could not have acted high up on the flanks of the mountains as they stand to-day. Deep beneath these tossed and crushed sediments enormous masses of rocks of entirely different origin—rocks which must have been crystallized slowly out of a highly heated magma—may often be found. It is a scene of confusion which might appear beyond human power to unravel.

In the plains, however, similar sediments are found. There they may exist in stratified beds arranged in a definite succession; and order begins to assert itself. They have, obviously, been deposited successively in the sea; the older beneath, the younger above. Further, as observations extend over the world, strata enclosing similar forms of fossil life appear and—more especially by the study of those finer-grained sediments which are of deep-water origin—it becomes possible to correlate the rocks according to their organic content. Such strata are of the same age. They have been formed simultaneously over the earth.[1] In this way, and in this way only, is a general chronology of terrestrial surface history rendered possible.

With this clue we return to the study of the mountains. All over the earth we find that these greatly displaced strata are, mainly, of comparatively recent date in the world's long history.

This is knowledge of great value, for it suggests that something happened which affected *simultaneously* the whole surface of the globe within recent times : something which caused the sedimentary beds of the preceding ages to become bent into folds and finally to be pushed up above the surface level of the earth.

We conclude that although this was comparatively recent it was a long while ago ; for we know that the forms of the mountains as they stand must be due to the slow process of denudation by rain and frost. In order that the pressures which folded them could have operated, the deep-cut valleys must once have been filled to the very brim and the whole buried beneath the earth's surface. And when we trace the over-folded strata from one part of the mountain range to another we find that vast thicknesses of formerly over-lying strata have been removed.

In the light of this great fact we return with larger ideas to the study of the earth. Our old ideas of 'antiquity' are gone. A new time-sense has developed in us, without which we cannot comprehend what is around us; we cannot reason about events which we are called upon to contemplate. Our time-units have become millions of years.

That the present is the key to the past is a fundamental

[1] We use the word 'simultaneous' in its geological sense as referring to a coincidence which a few scores of thousands of years does not affect.

## North Side of Grand Cañon, Colorado

In this wonderful dissection of the earlier sedimentary crust of the earth some of the oldest rocks of the geological sequence are represented—marked G. These are gneisses (altered sediments) of Archaean age. They are penetrated by granite intrusions. Great erosion has affected these rocks. The period of erosion was followed by sinking and submergence beneath the sea. During the submergence the Unkar and Chuar shales and limestones, some 12,000 ft. thick, were deposited—marked U. These are probably of Algonkian age; that is, they are some of the earliest rocks in which life has been recognized. Uplifting followed, accompanied with faulting and tilting, and the erosion which ensued swept away much of the Unkar and Chuar sediments. In some areas all was removed.

Another submergence then took place and the deposition of the Tinto sandstones and shales (marked T and Sh), to a depth of 800 ft. or more, followed. These are probably of late Cambrian age.

Ordovician, Silurian, and Devonian times are not recorded.

In early Carboniferous times there was again submergence and the Redwall limestones (marked R) were deposited to a thickness of 500 ft. or more. Upon the top of these, red muds and sands—the Supai sandstones, marked S—were laid down in shallower water to a depth of at least 1,000 ft. These are succeeded by the Coconino Beds (marked C) of shallow water character and about 300 ft. thick. After this the water again deepened and the Kaibab limestone (marked K) was deposited in later Carboniferous time. This was a slowly collecting deposit, and as some 700 ft. of it was formed it represents a very long time interval. This constitutes the present surface rock. There is evidence that many thousand feet of Permian, Mesozoic, and Cenozoic deposits have been removed by denudation in more recent times. (See *U. S. Geol. Survey, Bull.* 631, p. 126 *et seq.*)

North side of Grand Cañon; Colorado

law of Geological Science. If, then, the recent ages have seen the birth of great mountain ranges all over the earth, what about the remoter ages which must once have been? For in places here and there over the continents we find it possible to trace back the sediments through vistas of time so great that even the denudation of the young mountains covers a short period in comparison.

Going downwards, we find that scores of miles of sedimentary deposits have to be traversed and still life persists, although its complexity diminishes. Gradually the living world of our time disappears. Yet lower and the higher forms of life die out—vertebrate life at last becomes restricted to the seas. Then it, too, dies out. Continental life, air-breathing life, altogether goes, and we are left with the invertebrates and the vegetation of the ocean: corals and molluscs; brachiopods and crinoids; trilobites and graptolites; worms and algae.

The study of the rocks of remote past ages reveals the fact that mountain ranges were built up in those times also. True their worn-down stumps alone remain. What they were like in height and extent we can only very imperfectly realize. But that great ranges were in those remote times also fashioned simultaneously over the earth's surface can still be traced in their worn remnants.

Finally, we go back to times so remote that either life was not yet existent on the earth or its fossil remains have been completely obliterated, and even in those times great mountain-building periods seem to have affected the globe. The base-levelled stumps of Archaean mountains cover two millions of square miles in Canada, and in all parts of the earth where the younger rocks are removed we recognize the folded and metamorphosed remains of a primeval world.

We have begun our considerations at the climax of

Geological Science rather than at its beginning. But it has always been so in the history of Science: the fundamental facts are the last to be revealed. Recapitulation has to begin where research leaves off.

The subdivision of Geological History according to the lesser and more frequent events was accomplished by earlier workers in the science. Upon two criteria it was based—the lesser physical breaks revealed by the rocks, and the palaeontological changes.

Neither is in every case trustworthy, for each may be local and therefore deceptive. A well-known example is the existence both of a physical and a palaeontological break at the meeting of Carboniferous and Permian sediments in Great Britain so marked as to justify their systematic separation. But, as all recognize now, in many parts of the world the palaeontological break does not exist. We must, indeed, conclude from the very principles of organic evolution that a palaeontological break can be a local circumstance only.

Notwithstanding this objection to the 'systems' determined by a past generation of geologists, they have done good service to the science. Possibly one day they will be replaced by a more scientific time-scale. Meanwhile the first effort of the student of Stratigraphical Geology must be concentrated in acquiring some knowledge of the physical and biological characters of the successive systems. He must, however, bear in mind that these systems lay no claim to equality in duration. It is often said that the older systems represent longer durations than the more recent systems. It is probably true that the fine sediments which build up Cambrian, Ordovician, or Silurian deposits represent a longer time-period than equal thicknesses of Devonian sandstones, Carboniferous coal-measures, or Jurassic limestones. And the fact that the fine-grained, slowly-collecting sediments are often especially abundant in the older systems goes far to support the claim that these indeed represent very long time-periods. But at present little more can be said.

In comparing recent with past geological history the geologist is under a considerable disadvantage. He has to contend with the denudative destruction and metamorphic changes of the scores or hundreds of millions of years between then and now. How, for instance, is he to determine the existence of continental conditions as having prevailed locally in some past age? Continental conditions enduring for millions of years may leave little behind save loose surface deposits or shallow clays laid down in fresh-water lakes, possibly coal-measures, possibly signs of glaciation. True these may be quite distinctive when preserved for his observation; but they are readily removed when continental conditions are coming to a close and the sea is advancing upon the land. The only indication remaining in such a case may be an unconformity. An unconformity comes into existence when rocks which have become worn down by denudation are then submerged so that fresh deposits are laid down upon the worn surface. Subsequently, if again uplifted, the denuded meeting plane of the older and the newer strata tells the geologist that continental conditions must have existed; for only under continental conditions does denudation occur. Denudation tells him of former continental conditions just as deposition tells him of sub-aqueous. Thus the unconformity may be a clue to former emergence of the land above the sea.

Other conditions attended by great compressional and tensional forces, long ago prevailing in the crust, have to be deciphered. Compressive conditions may be revealed by folding or, in extreme cases, by the overturning of the folds; the arched-up anticlines being bodily displaced, thrown down, and even faulted in such a way that the upper limb of the horizontally-extended fold is displaced over the lower limb. Such displacements are, in mountain regions, sometimes seen to involve movements of many miles.

Faulting reveals compressional conditions or tensional conditions. Compressional faults show reduction, tensional faults show increase of the original horizontal extension. Rifting also reveals former tensile conditions. In some parts of the globe rifts are developed on a very great scale, as has been more fully referred to in the first chapter.

The determination of the direction from which a thrust has come is one of the fundamentally important problems facing the geologist. The phenomenon already referred to, the faulted and over-thrust anticline, is one clue to its solution. Again it will often be found that the deformation increases towards the direction from which the thrust has come, or that older and deeper-lying beds have been brought up in this direction. On the whole the amount of disturbance increases towards the source of pressure, whether it be shown in faulting, or folding, or in the alteration of the rocks by heat and pressure, i. e. metamorphic changes. This last criterion is of much importance, for it is often difficult to distinguish between the effects of under-thrusts (i. e. deep-seated thrusts) from one direction and over-thrusts (i. e. thrusts higher up) in the opposite direction. Both may produce overturned folds leaning in the same direction.

Some of the most striking and important facts of Geological Science arise out of the study of the mountains. Later we shall specially refer to this subject. It is enough to say here that the mountain ranges of the earth have been built up out of folded sediments. The moun-

Two views of the Matterhorn (14,780 ft.)

The upper view is from the summit of the Zinal Rothhorn (13,855 ft.) and shows the northern face of the mountain. The lower, from the Mettelhorn (11,188 ft.), shows more of the eastern face. The Matterhorn is believed to be weathered out of one of the great recumbent folds of the Swiss Alps. It is, according to this view, a mountain without roots and foreign to its surroundings. The contortions of the hard gneissic and schistose rocks which enter into its composition testify to the extreme dynamic conditions to which it has been exposed.

tains have, in fact, risen from the sea-floor. In the heights of the Himalayas are rocks which were slowly deposited in the depths of the seas in comparatively recent times.

This fact gives us the clue to the age of the mountains. Clearly the Himalayas must have been uplifted *after* those beds had been formed. And we have to recognize that where the mountain now stands the sea had once stretched in pre-historic solitude. Age-long accumulations of sediment—to depths of many thousands of feet—may have been formed before the great resurrection took place. Evidently such great depths of sediment could not have accumulated if the floor of the ancient sea had not been sinking. Because of their great extent such vast subsiding areas are known as geosynclines. The geosyncline wherein the Cordilleras of N. America were cradled extended from the Gulf of Mexico to the Arctic Ocean and was more than 1,000 miles in width.

The surface history of the earth has been by no means uniform and monotonous. Great physical changes have been repeated at intervals. It would appear as if six great cycles of world-transforming events are recognizable during the course of geological history. In each of these cycles the succession of events has been the same. The continents sink relatively to the ocean. The waters flow in over the lower levels and vast areas become covered by the transgressional seas. These seas persist over very long periods—fluctuate in area—advance and retreat, often many times—but always still advancing until at length a time is reached when retreat overtakes advance and little by little the land rises again. And now a strange climax is attained. Just where the seas have been most enduring, mountains begin to arise. First it seems as if lateral forces were at work. For the rising deposits—the age-long sedimentary accumulations of the geosynclines—come up crushed, folded, and even over-thrust for many miles. These are destined to form the mountain ranges of the ensuing era. A last great vertical uplift, long after the first deformation of the sediments, raises these new-born mountain ranges high above the continental level: an uplift which may amount to many thousands of feet.

Then succeeds comparative repose. Evidence of cold climatic conditions often attends the period of greatest continental elevation. These conditions finally pass away after some thousands of years, telling of renewed sinking of the land. And this period of very slow sinking endures over millions of years, approximating ever more and more to the time when once more the seas shall flood the continents. And so the cycle of events begins all over again.

This extraordinary history is no myth. It has been traced in many parts of the world. There is sufficient agreement as to the spacing of these majestic events to assure us that in its broad outlines the same succession of changes has affected the entire surface of the earth.

We live to-day in the early stages succeeding a great epoch of mountain-building. All the greatest ranges of the earth are of comparatively recent date. The Alps, the Himalayas, the Caucasus, the Pyrenees, the Andes, the Rockies, &c., all are newly-erected features of the globe. True, these ranges may be in a sense but a rejuvenation of more ancient ranges which time had all but demolished: the buried mountains, out-worn things, being hidden far beneath the snow-clad peaks which to-day rise above the clouds.

And brief, geologically brief, as the period has been since the last great resurrection of the mountains, see how time has already dealt with them! It would appear as if more than half their original, tremendous bulk of

The Dent Blanche (14,318 ft.) left, and the Grand Cornier (13,023 ft.) right, are separated by the beautiful Dent Blanche Glacier

The view is from the summit of the Zinal Rothhorn (13,855 ft.) and looks to the south-west. Beyond, in the further distance, is a tumultuous ocean of mountains, among which Montblanc is the central and furthest object.

hard and durable rocks had been hurried away: carried into the ocean or spread over the lowlands.

There are no transgressional seas now upon the earth to conserve these remains upon the continents. But we know that the invading seas must surely come again and fresh geosynclines must develop in which the great interment must take place.

Attending these mountain-building or orogenic periods volcanicity may develop on a great scale, and, during the period of down-sinking of the continents, fissures may appear in the lower levels or along the ocean margins, through which stupendous volumes of basalt may flood out over the land or over the sea-floor—such floods as we have referred to in the first chapter.

American geologists designate the mountain-building periods 'revolutions'. Looking back, we have to recognize that the revolutions and the long preparatory periods leading up to them constitute the whole of earth-history. One revolution has succeeded another. Possibly the whole series is not yet deciphered. There are minor fluctuations of level, minor uplifts, minor orogeneses, at present regarded as of local origin. It may turn out that some of these will have to be recognized as world revolutions which, although involving age-long preparations, may yet be less deformative in their effects. To these possibilities we must refer later. At present we can only definitely accept, on the authority of those most competent to guide us, that some six or seven great revolutions have dominated earth-history since the beginning of geological time.

We give them here in their successive order and place among the geological systems. Later they will be discussed in more detail. In each case it is the mountain-building period which is referred to as the revolution.

# THE REVOLUTIONS AND DIVISIONS OF GEOLOGICAL TIME

| *Period.* | *System.* | *Life.* | *Revolutions.* |
| --- | --- | --- | --- |
| | Recent | Dominance of Man. | |
| | Pleistocene | Palaeolithic Man. | |
| Tertiary (Kainozoic) | Pliocene | Advent of Man? Highest orders of mammals and plants. | |
| | Miocene | Continued advance towards recent forms—especially in molluscan and insect life. | Alpine. |
| | Oligocene | Continued advance towards recent forms. | |
| | Eocene | Dawn of recent forms of life. A marked advance over Cretaceous forms of life, especially in the Mammalia, is found in earliest Eocene. Early angiosperms. | |
| Secondary (Mesozoic) | Cretaceous<br>Jurassic<br>Triassic | 'Age of Reptiles'—(both herbivorous and carnivorous)—on land and in the sea; prophetic, in their various forms, of birds and mammals.<br>Remarkable evolution of gastropods, cephalopods (ammonites), and bivalves, advent of Mammalia. Early cycads and conifers.<br>Disappearance of Palaeozoic seed ferns, Cordaitales, and Lepidodendron. | Laramide. |
| Primary (Palaeozoic) | Permian | Evolution of air-breathing, vertebrate life continued. Advance in insect and plant life. Trilobites disappear. | Appalachian. |
| | Carboniferous | Great development of fern-like plants and of insect life.<br>Ancestral amphibians on the land.<br>Bivalve, crinoid, and coral marine life abundant. | |
| | Devonian | 'Age of Fishes'—(armoured and enamel-scaled types); first land floras; precursors of the amphibians; marine invertebrate life abundant, especially molluscs; brachiopods; corals. Decline of trilobites. | |
| | Silurian | Fishes rare at first; later abundant. Life mainly represented by corals, brachiopods, trilobites, crinoids, bryozoans, and graptolites: the last become extinct in Silurian times. | Caledonian. |
| | Ordovician | Advent of true corals and armoured fishes. Rise of shelled marine life (lamellibranchs and brachiopods); bryozoans and graptolites. | |
| | Cambrian | Dominance of trilobites; rise of cephalopods; primitive corals and sponges; brachiopods abundant; early lamellibranchs and crustaceans; land plants and land animals unknown. First known marine faunas. | |
| Proterozoic | Keweenawan<br>Huronian | Worms; radiolaria; siliceous sponges. | Killarney. |
| | Timiskamian | Calcareous algae. | Algoman. |
| Archaean | Loganian | No trace of life. | Laurentian. |

## VI

# THE SOURCE OF THE REVOLUTIONS

In foregoing chapters we found that the structure of the earth's surface revealed continents floating in a vast and universal layer of basaltic composition, both continental rocks and the basaltic magma being feebly radioactive. Where the substratum is uncovered the oceans rest upon its surface. We found that at the present time the substratum is in the solid state: and, as it cannot appreciably lose heat by conductivity through the continents and only to a very limited extent and very slowly by conductivity to the oceans, the radioactive heat continually evolved throughout its mass must accumulate in its entirety.

Reviewing the surface history of the earth, we then found a record of periodic flooding of the continents, followed after long periods of time by retreat of the transgressional seas, and then by a great epoch of mountain-building and volcanism; succeeded finally by quiescence and such conditions as now prevail, wherein the land is above sea-level and the only surface activities are those of rain and rivers, frost and thaw. Thus a cycle is completed but only to be merged into a succeeding cycle in which the seas once more gradually steal in over the lower continental levels.

Can we trace the source of these great cyclical events—each cycle involving some scores of millions of years for its completion—to the physical constitution of the earth's surface?

We have already seen good reason for believing that the substratum, although solid, is near its melting-point. Further evidence will presently arise: meanwhile we assume that this is indeed the thermal state of the substratum, and looking forward into the future we shall inquire into the results arising out of the ceaselessly accumulating radioactive heat.

Recent examination of basalts of deep-seated origin, i.e. those which have been poured out in great floods on the continental surface, affords for the radium content $1 \cdot 0 \times 10^{-12}$ grm. per grm. and for the thorium content $0 \cdot 8 \times 10^{-5}$ grm. per grm. These figures are the mean of those derived from three areas: the Deccan, the Hebridean, and the Oregonean (of Western N. America). The general mean is raised somewhat by the radioactivity of the Oregonean basalt. Multiplying by $5 \cdot 6 \times 10^{-2}$ in the case of the radium and by $6 \cdot 6 \times 10^{-9}$ in the case of thorium,[1] we find that the heat produced per second per grm. of basalt is $0 \cdot 11 \times 10^{-12}$ calorie. If the density be $3 \cdot 0$, the evolution per c.c. is $0 \cdot 33 \times 10^{-12}$ calorie per second.

We shall now apply this result to the substratum. It is already known to the reader that a substance which is at its melting-point but still retains the solid state will liquefy if supplied with its latent heat of liquefaction. The value of this latent heat for basaltic rocks has been investigated by Vogt, who finds the latent heat to be about 90 calories per grm.[2] Hempel and Heraeus find for the latent heat of the mineral melilite (analogous to basalt in chemical composition) 90 calories. Doelter states that the latent heats of fusion of silicates approximate to 100 calories per grm.

But we saw in Chapter III (p. 53) that something more than the latent heat has to be supplied. When the basalt last cooled from fusion it fell in temperature some 100 degrees below the melting-point before crystallizing.

[1] See Chap. IV.
[2] *Christiania Vid. Selsk. Skriften*, 1904, pp. 54 et seq.

There was then a sharp increase of density and a rise in temperature probably due to the liberated latent heat of solidification. We shall neglect this source of heat and start with the definite temperature of 1,050° C. We have then to supply first such a quantity of heat as will bring the temperature of the basalt to its melting-point—i.e. to 1,150°.

This quantity depends on the specific heat of the basalt over this range of temperature. Now the specific heat of substances of this kind—i. e. the silicates—increases with the temperature; but the increase generally ceases at about 500° C. In the case of the rock-forming minerals it is then about 0·23 calorie; that is, it will take this quantity of heat to raise the temperature of one gram of the rock by one degree. Hence, to bring the basalt to 1,150° requires $100 \times 0{\cdot}23 = 23$ calories per grm.

The latent heat required to change the state from solid to liquid has next to be supplied. The latent heat may be taken as 90 calories. So it requires in all 113 calories to bring a single gram of the rock into the fluid state.

The foregoing data have been derived from measurements made at atmospheric pressures. That they will be modified somewhat at such pressures as prevail beneath the continents is certain. The influence of pressure on the latent heat will probably not be very great. The energy here is mainly expended on internal work, and this part is not likely to be seriously affected. That part which is expended on external work will, however, be increased. It seems, on the whole, probable that observed values of latent heat are not very much affected by the conditions prevailing in the upper parts of the magma.

Now the rate of thermal evolution found above for basalt, i. e. $0{\cdot}11 \times 10^{-12}$ calorie per grm., amounts to 3·46 calories per grm. in one million years. We see, therefore, that about thirty-three million years must elapse in order that the requisite heat may accumulate and fusion be brought about.

We must regard this result as an approximation only. It may be that the radioactivity determined for the Deccan and Hebridean basalts more nearly represents the average radioactive state of the substratum. In this case 2·2 calories per grm. collect in one million years, and a period of about fifty-six million years would be required (on the larger estimate of the latent heat) to bring about fluidity.

In Chapter III we have already considered the effects of a change of state of the substratum upon the buoyancy of the continents, and we arrived at the conclusion that the laboratory results respecting the density-change must be considerably affected by the pressure conditions prevailing beneath the continents: a conclusion to be anticipated from our knowledge of the effects of pressure on the density of fluids.

It seems probable that the break-down of solidity begins in the upper levels, and the change of state progresses gradually downwards. Ultimately the fluid state will come to prevail around even the deepest compensations of the raised surface features of the continents, which the principle of flotation calls for.

Consider now the compensations dipping, in some cases, many miles into the substratum, which at this time has lost an appreciable part of its density. These compensations no longer possess their former sustaining power. Some part of the support of the continents is withdrawn. This effect is equivalent to a great load being imposed upon the continental surface. It must sink deeper into the magma. This is quite inevitable.

As the ocean floor melts away beneath, it leaves the continental layer ever more and more exposed to the fluid magma, so that it ever more and more is affected by the loss in the density of the substratum and, in obedience to isostasy, sinks deeper.

It is true that the general increase in the volume of the substratum causes the earth's radius to augment by a little—a very few miles—so that continents and oceans are everywhere borne upwards by this few miles. But this does not interfere with the down-sinking of the continents in the substratum.

The oceans do not experience this effect, for they simply repose upon the upper surface of the basaltic ocean floor, and this floor of cooled basalt possesses no great compensation projecting downwards into the liquefied magma.

Thus the general primary effect will be a slow and very gradual sinking of the continents relative to the ocean level, so that the waters of the ocean must transgress upon the lower levels of the continents. In short, forces are brought into operation which affect the continents but do not in the same manner affect the oceans. These forces result in vertical displacement of the continents relatively to the ocean floor.

Consider the conditions prevailing at this period over the earth. The solid crust of the earth, consisting of the continents and ocean floor, are being stressed by the increasing volume of the substratum. The earth is, in fact, increasing in volume, and its solid crust is too small to fit a larger world. Two effects inevitably follow—fluid pressure in the substratum and corresponding tensile stresses in the crust. From these conditions what effects may be expected to arise?

The ocean floor will, mainly, take up the increase in the surface of the globe. This follows from the fact that at this time it is being reduced in thickness and is, probably, considerably less in thickness than the continental crust. We expect the development of rifts in the stretched surface crust, and more especially over the ocean floor and along the continental margins where special differential stresses arise: shearing forces tending to part the continents from the basaltic floor of the ocean. Through such rifts the melted magma will be forced at high pressure. It will flood out over the sea-floor where rifts have developed therein and great dikes have been formed. Such lava will soon re-solidify under the cooling effects of the ocean. Along such rifts many of the Pacific islands would seem to have been elevated, as we saw in Chapter I. Where the lava flows out along the continental margin we find such resulting surface features as we see to-day along the coasts of North-Western Europe—the basaltic coastal features of Northern Ireland, Western Scotland, &c. Again, sometimes these fractures will release the imprisoned magma upon the land, as happened in the out-flows of the Deccan Traps in Western India. Nor will the continents escape the tensile forces. Rifting will surely occur and, as we have already seen, is very much in evidence.

It must be kept in mind that this down-sinking and flooding of the continents will be a very slow process and must take many millions of years to accomplish. At the present time the continents are highly elevated because the substratum is solid and possesses its greatest density, but, as we have seen, the latter must be steadily accumulating heat and gradually changing state. The ultimate sinking and flooding of the continents is a certainty, but it will take place at a very remote period in the future.

When the liquefaction of the upper region of the substratum has progressed a certain way, a new factor intervenes—the tidal effects arising out of the gravitational influence of moon and sun. These effects have already been referred to as necessarily attending the existence of a fluid substratum beneath the continental crust. The disturbance at the surface would probably be slight. The ocean tides would to a large extent disappear, being in great part replaced by a tidal rise and fall of the whole outer solid crust of the earth.

But although the surface effects, as observed by intelligent beings living at the time, would be small, the consequences of this tidal effect would be very important and indeed may be regarded as probably necessary for the habitability of the globe by the higher forms of life. For, in fact, mathematical investigation has shown that the tidal forces involve a slow shifting of the entire outer crust—continents and sea-floor and the sustained oceans—over the fluid magma beneath: the motion relatively to the globe within being westerly. In order to see the importance of this we must enter on the question of how the accumulated radioactive heat ultimately escapes from the substratum.

The escape of heat must occur mainly through the ocean floor. For although some heat may (and probably does) flow upwards through the continental layer, this means of escape cannot relieve the accumulation within. It leads doubtless to the freezing of a shallow layer of basalt over certain areas of the continental base where the crust is thin. But the vast magmatic region beneath must be isolated and cut off from this feeble leakage. Hence there is a grave risk of superheating beneath the continents. If relief did not arise we can only suppose that catastrophic igneous effects must ultimately ensue—possibly competent to destroy subaërial life on the globe.

But the tidal shifting movement of the outer crust, which automatically arises when a certain amount of heat has collected in the substratum, averts this evil. For the westerly drift of the crust involves, of course, the shifting of the ocean floor over areas whereon, in prior ages, the continents had reposed. And as the slow movement continues, by this beautiful mechanism the cooling influence of the oceans is carried over the surface of the substratum. Every part of this surface is exposed to the same condition. To every part the same means of thermal escape is extended.

We must now briefly consider the origin and growth of the ocean floor. By the ocean floor we mean that part of the substratum which bears the ocean upon its upper surface and which is throughout at a lower temperature than that of the melting-point of the magma. Its temperature augments downwards, and only at its base ultimately attains the melting-point. If the great interval of slow accumulation of radioactive heat lasts long enough this floor grows to such a thickness that its basal temperature *due to its own proper radioactivity* is the melting-point of the magma. It can grow no thicker. Possessed of this limiting thickness, it necessarily assumes the same heat-conserving property we recognize in the continental layer. Thereafter heat from beneath cannot escape upwards for exactly the same reason as it cannot pass upwards through the continental layer. If, on the other hand, there has not been time for this amount of cooling to occur there will still be some escape from the underlying magma through the ocean floor. The matter has been mathematically discussed by Cotter.[1] He finds that in thirty millions of years—i. e. in

[1] *Phil. Mag.*, September 1924.

the interval of thermal accumulation according to the higher value of the basaltic radioactivity—the base of the ocean floor attains a depth which is greater than 19·6 and less than 26·2 miles, and that the limiting thickness of the ocean floor is about 30 miles.

When at length liquefaction begins to prevail in the substratum, this floor is rapidly attacked by the circulating magma: some part of which, rising from beneath, is superheated respecting the conditions at higher levels.[1] Much of this floor is, of course, near the melting-point. A relatively short period must see it greatly reduced in thickness. Thus if the melting away of the floor was only 10 cm. in a year, its reduction from 48 km. (30 miles) to 6 km. (4 miles) thickness would take less than half a million years.

It seems probable that reduction in the thickness of the ocean floor cannot proceed beyond a certain point; that there will be a minor limit to its thickness, i.e. that at which the rate of thermal escape is just equal to the accessions of heat from convective movements. But, of course, a numerical estimate of such a limiting thickness is not practicable.

We can, however, form some idea of what goes on in the substratum during this period of heat-loss. For one thing, it seems certain that re-solidification, although commencing above, will build upwards from beneath. The reason for this is not far to seek. Re-solidified magma in the upper reaches assumes a higher density than the surrounding fluid magma. It accordingly sinks, and as it descends it is ever exposed to increased pressure tending to increase its density. It is probable that heavier and more ferruginous lava lies beneath. The lighter magma is here arrested. We note that the down-

[1] See Appendix to this chapter.

ward movement of course entails the upward movement of corresponding volumes of magma.

But in addition to these convective movements the tidal disturbance travels twice diurnally round the earth. Lava has to flow in to feed the tidal elevation, and this must be a potent source of disturbance, continuing till tidal motions begin to die out with the advance of congelation.

We can now see how absolutely essential to the discharge of the radioactive heat is this influence of the tide-generating forces—an influence which in the past, if Darwin's lunar theory is correct, must have been greater than it is in our time. Look at a globe of the earth and imagine that some force, acting approximately parallel with the equator, is pulling the outer surface of the globe —oceans and continents—from right to left—i. e. westward. You will see that owing to the manner in which land and water are distributed on the earth's surface the ocean must travel over the places where the continents had previously been.

The oceans rapidly absorb the heat which has accumulated over thirty to fifty millions of years. Casual consideration might suggest the idea that here we might possess a source of climatic change—of warm climate during periods of cooling of the magma, of colder climates during periods of accumulation. But it is not so. It can be readily shown that at no time, and under no possible assumptions as to the rate of thermal discharge, will there be any appreciable warming of the ocean.

A simple calculation will suffice to dispose of this suggestion. On the data necessarily arising out of the conditions, the rate at which heat can pass through the ocean floor depends on the thickness of this floor; that is, if we assume that the supply from beneath by convective currents is adequate to maintain the temperature at the base of the floor at the melting-point. As already stated, there must be a limitation to the thickness of the floor arising out of this last condition. Let us assume that the floor—after an initial period, during which most of the heat imparted to it is expended on supplying the latent heat necessary for its reduction—finally diminishes to a thickness of 6 km. (3·8 miles). We assume that the conductivity is $4 \times 10^{-3}$, and that the gradient is, say, $1,200^\circ/6 \times 10^5 = 0^\circ{\cdot}0020$. Multiplying by the conductivity, we find that $80 \times 10^{-7}$ calorie would pass through per second per square centimetre—an absolutely negligible quantity. And it must obviously remain negligible whatever reasonable assumption we may make as to the ultimate thickness of the ocean floor or the rate of thermal loss by effusion of lava into the ocean.

As the heat escapes and the magma solidifies from beneath upwards the tidal movements must die out; very gradually, doubtless. The deeper compensations are the first to be involved in the slowing down of tidal drift. Finally, the drift movement ceases or dwindles to its present almost insensible amount.

We may attempt a rough estimate of the time required for the heat taking part in a revolution, and which has accumulated over a period of from thirty to fifty millions of years, to escape into the ocean. Tidal circulation in the magma is supposed to be in progress. Let us limit the depth to 32 km. (20 miles) beneath the continents. Our thermal calculations are simple. The ocean floor is, say, 6 km. thick and, as already found, it passes out $80 \times 10^{-7}$ cal. per second or 248 cals. per year per square centimetre. We shall deal with a vertical column one square centimetre in cross-section extending all the way down. Now the heat stored in 32 km., where the latent heat is 90 cals. per grm., or 270 per cubic centimetre, is evidently $32 \times 10^5 \times 270 = 864 \times 10^6$ cal., and, dividing by 248, we may take it that 3·5 million years will suffice to convey away all the heat. We may include all that has collected beneath the continents down to the same depth by adding 2/5ths, making the period required nearly $5 \times 10^6$ years. The time required for the reduction in thickness of the ocean floor is not included.

It is not possible to determine whether such estimates may or may not be applicable to considerably greater depths. All we are fairly sure of is that the rate of accumulation of heat per grm. is everywhere much alike. But the physical conditions attending its storage are very imperfectly known to us. And, of course, the thickness assumed for the ocean floor is arbitrary. It may be more or less reduced according to the intensity and duration of a revolution.

Again, there must be very considerable thermal loss arising out of effusive discharge of lava into the ocean. Such loss *must* occur and upon a very great scale, because, in fact, the floor is subjected to breaking stresses concurrently with and arising out of magmatic pressure beneath. The ocean floor undoubtedly takes up the greater part of the areal extension of the earth's surface and has to accommodate an increase of area which in the long period of crustal expansion must add up to several hundreds of thousands of square miles.[1] Hence calculations of thermal loss based upon conductivity must exaggerate the period required for the discharge of accumulated heat.

What happens at the surface during this epoch of heat-loss? As the magmatic density is restored the earth's radius gradually recovers its original value. Both conti-

[1] See Chap. VII, p. 115.

nents and ocean floor all over the earth sink downwards. But obviously something else is involved. The increased density of the magma floating the continents causes these to rise and to float at their former level. Again the effect here is first upon the deeper compensations. The continents rise relatively to the oceans and the transgressional seas must disappear from the land and flow back into the ocean.

The cycle is now complete. The substratum is once more solid. The land stands high above the sea-level and the long period of accumulating radioactive heat is beginning all over again—a period so long that in its passage the phantasmagoria of life and death ushers in undreamt-of forms upon the earth.

Events take place at the surface to which we must now refer. One of these may be briefly dismissed. The temporary increase of the terrestrial radius involves a slowing-down of the earth's axial rotation. This increase of radius is now at an end. The former rotational rate is restored save in so far as the gravitational interaction between moon, sun, and earth may have pursued its dynamical history. These changes in the length of the day would appear to affect but little the matters we are discussing.

The reduction of the terrestrial radius involves, however, effects which do very much concern the surface history of the earth. For this change is attended by a reversal of the former conditions of tension in the outer crust of the globe. It is now thrown into a state of pressure, and this condition must prevail not only in the ocean floor but in the continental crust. As it is very probable—indeed certain—that the previous enlargement of the terrestrial crust mainly took effect upon the ocean floor, so now we may regard the major compressive stresses as arising from an ocean floor grown too large for its former place upon the globe. And here we are confronted with the principal source of orogenesis on the face of the earth.

In this chapter we have now traced the origin of the crust movements which culminate in the 'Revolution'. The latter name is applied by American geologists, not to the whole sequence of cyclical events, but to the great events attending the folding and elevation of the mountains. Having traced the physical events up to those surface-movements which develop into mountain-building, we, of necessity, leave off till we have before us a more detailed account of the surface phenomena which attend a revolution and for which we have to find explanation.

We shall conclude by a brief consideration of the question as to whether some other sequence of events than that which is herein described might not have arisen from the fundamental facts of the existence of isostasy and a radioactive substratum.

This question resolves itself into another and more definite one: Are the storage of heat by the substratum and its periodic escape inevitable events? For, in fact, these are at the basis of all the geological phenomena we have been discussing: the sinking of the continents, their re-elevation, and the development of the crustal stresses leading to mountain-building.

Let us start from the present state of the substratum. It is now solid. But if it is solid, then the radioàctive heat continually being developed throughout every part of it cannot escape; or only very partially. For beneath the continents it is almost entirely accumulated. Beneath the ocean it can only escape by conductivity, and as the evolved ocean floor grows thicker, here, also, must ultimately accumulate all heat generated in the depths. *Hence the present condition of solidity cannot continue. Melting must ultimately come about.*

But we also know that the fluid state then arising cannot be permanent. We know this on physical grounds. We may also reason that if it were then the substratum would not now be solid. And hence as the fluid state which we have proved to be inevitable is not permanent there must arise conditions permitting the stored heat to escape and the solid state (the present one) to be restored.

Hence we see that the existing conditions involve periodic physical changes in the state of the substratum from solid to liquid and again back to the solid state, and so on.

If we look round us to see whether we can evade this conclusion we find that we can only do so by contending that the existing condition is, in fact, not a conservative one : that heat *is* escaping in some manner as fast as it is being generated throughout the substratum. But how ? It cannot be by conductivity. The blanketing effects of continental radioactivity and of the radioactivity of the ocean floor effectively block its escape by conductivity. Our only resource then is convection. Dikes, innumerable, must extend downwards beneath the ocean floor to the ultimate base of the substratum, and in these dikes (probably *not less* than a hundred miles in depth) melted magma must circulate ; ascending currents yielding their heat to the ocean and in some way returning in order to preserve the circulation. But the difficulties which confront us respecting sub-oceanic regions are intensified when we turn our attention to the sub-continental conditions. Beneath the continents we can only appeal to horizontal sills which must penetrate many thousands of miles in every direction. What force can maintain the horizontal circulation? It cannot be gravitational. Clearly these hypothetical conditions are impossible.

Apart from these difficulties a medium so intermixed with fluid matter as must be postulated would not satisfy the requirements of seismology. It would not transmit distortional seismic waves.

Considering this matter, we see that the source of the periodicity lies in the facts that thermal discharge can only take place by convective movements and the co-operation of tidal effects; and that the rate of loss arising in this way so enormously outstrips that of the extremely slow rate of supply that reversion to solidity becomes inevitable: an important condition being the growth of solidity from beneath upwards.

The paroxysms of the geyser or the intermittent activity of volcanoes doubtless operate on similar principles.

Finally, even if a physically tenable alternative explanation were forthcoming, it must fail if it cannot meet the requirements of geological science—in other words, if it cannot account for the surface phenomena known to us. Now no imaginable physical system which results in conditions of steady heat-flow to the surface can account for those periodical events which are at the basis of geological history and which have hitherto constituted its most elusive problem.

## APPENDIX TO CHAPTER VI

### *Physical Conditions prevailing in the Substratum*

As stated above, we know very little about the behaviour of highly heated rocks at great pressure. It is known on theoretical principles as well as experimentally that the melting-point of a substance which expands on liquefying (as basalt does) is raised by pressure. In the depths of the substratum the melting-point is therefore higher than we observe it to be in the laboratory. There are reasons for believing that the increase would reach a maximum at a depth of about 150 km. (90 miles). The conclusions reached by Vogt would entail a rise of melting-point of 50° C. for a depth of 25 miles. Thus, basalt which melted at 1,150° at the surface would melt at 1,200° 25 miles down—say at the base of the continents.

When circulation commences in the substratum, this would give rise to upward currents of superheated lava. For the deep-lying lava which has been liquefied must be at a higher temperature than lava liquefied at lesser depths. If then it is carried up by tidal disturbances it will be very effective in supplying the heat necessary to reduce the thickness of the ocean floor. On the other hand, when lava becomes congealed in the upper layers and, in virtue of its increased density, sinks downwards till arrested by such high-density magma as may exist below, it finds itself below the prevailing temperature (the melting-point being higher in the depths) and must absorb heat from its surroundings.

A further condition which must result in delaying liquefaction in the depths of the substratum arises from the increase of viscosity downwards attendant on increase of pressure. This effect may be small, but it probably exists.

## *The Ocean Floor and the Congealing of the Substratum*

In periods of liquefaction and tidal movements the floor must be at first rapidly reduced in thickness. For, in fact, much of it is near the melting-point, and currents from beneath are more or less superheated. The fusion of the floor must involve a considerable expenditure of heat, and this cannot come from heat which is exactly at the temperature of fusion. It is possible—but not probable—that before the floor is effectively attacked there is some general elevation of temperature above the actual melting-point. This may delay the advent of revolution. But much of the sub-continental magma, as well as such magma as is stirred up from deeper layers, must reach the ocean floor in a super-heated state.

It seems certain that the ocean floor cannot be indefinitely reduced in thickness. There will be a certain limiting thickness which will be attained when the rate of escape of heat into the ocean becomes equal to the rate of supply from beneath. It must be remembered that the rate of conducted heat-flow through the floor depends simply on its thickness, the temperatures of its upper and lower surface remaining almost constant. Now the rate of supply depends on the activity of convective movements. This must diminish as time goes on. When equilibrium is reached no part of the thermal supply is available for further reduction of the ocean floor, all passing through into the ocean either by conduction or by effusive discharge of the molten magma.

After a lapse of time during which equilibrium between the rate of escape and the rate of supply of heat from the deeper parts of the substratum exists, the supply must dwindle and a period of regrowth of the floor must succeed.

Tidal movements will die out slowly, and their ultimate cessation may leave some of the substratum still fluid. Beneath the oceans this involves the downward growth of the ocean floor. Beneath the continents there may be, locally, some further loss by conductivity through the continental crust. But congelation at all levels will be promoted by final thermal adjustment between solidified lava precipitated from higher levels and deeper-lying magmas possessing the melting temperatures proper to their horizons.

### *Stability of the Ocean Floor*

The question of the stability of the ocean floor during the period of liquefaction naturally arises. Will the floor break up and founder ?

The higher specific gravity of the plateau basalts contrasted with that of the volcanic basalts (see Chap. I, App.), and the effect of pressure upon the density of the underlying fluid magma, render it probable that a balance of buoyancy would remain to the 'floor' even if this were free to sink. But in any case it is obvious that the system of ocean floor interposed between cold ocean water of low density and hot liquid basalt is really a stable one. If the floor were to rupture and, under the action of stresses, were in part forced under, it could only be absorbed and re-melted by its own mother liquor, which it cools in the act and which the great heat-capacity of the cold water above must rapidly re-solidify. Such an event can therefore lead to no notable consequences save a more rapid loss of heat from the substratum.

But, in truth, foundering of this sort can only rarely happen, for cracks, whether produced by tension or by compression, must soon heal by the injection from beneath of lava under pressure. Such ruptures, doubtless, led to the outflow of the Hebridean basalt, and such ruptures were probably frequent during geological history, and have taken place on an enormous scale over the floor of the ocean.

The pressure at the bottom of the ocean is too great to permit of ebullition. The critical pressure, above which ebullition cannot take place, is about 200 atmospheres ; that prevailing at the average depth fully 400. Hence the heat would enter the water quietly and be rapidly conveyed away by convective circulation.

## VII

## THE BUILDING OF THE MOUNTAINS

When, after a very long period, the escape into the ocean of the latent heat stored in the substratum begins, the previous sequence of events becomes reversed. Shrinkage of the earth's surface sets in and the tensile stresses in the crust gradually change into compressive stresses. The outer crust of the earth is now too great for the diminishing surface area, and the sea-floor begins to bear against the continental margin with a force which augments as it grows in thickness and with the progress of the solidification of the substratum.

The forces so developed must be very great. The ocean floor is itself thrown into long undulations: such gentle slopes as give rise to the central shallowing of the Atlantic, or such steeper slopes as descend into the deeps of the Pacific. The coasts are crushed inwards, and raised borderlands are formed and mountain elevation inaugurated.

The most striking fact known about the mountains is that they are largely and often mainly composed of sedimentary rocks, that is, of rocks which had been deposited originally in the seas. True, these sediments may be contorted, folded, even metamorphosed almost beyond recognition, but none the less they have risen from the sea-floor to form the mountain chain. It is a universal fact. Even of the volcano-topped Andes and Caucasus it is true. In the great precipices of the Alpine heights the limestones, folded as if made of wax, reveal themselves in giant arches. The hard slates (often changed to mica-schist) telling of former deep waters, buttress such giants as the Eiger or the Matterhorn, and overlie or intermingle with the granites of the Himalayas. Why is this? We shall see that the clue to the origin of mountain ranges must be sought for in the eloquent fact of their marine origin.

Now the mountain chains have not risen from the central ocean, nor have we reason to believe that they arose off the coasts of a continent. They arose from inland seas or mediterraneans, generally elongated parallel with the continental coast from which they were many miles removed. Observe on the map the location of the Cordilleras of North and South America and the location of the Eurasian chains. The former trend more or less north and south; the latter east and west; the continental boundaries having similar trends. In North America there are two successive mountain systems (the Rockies and the Sierra Nevada and Coast Ranges) parallel with the western coast. In South America the Cordilleras consist of compacted ranges of different ages and structure. The curved ranges of Eurasia are mainly parallel with the southern continental margin. Looking at these ranges, we must recognize that the lateral forces which raise the mountains may operate far within the continental coasts. There is nothing surprising in this, for in fact the compressional stress must be conveyed far into the continent, and if there exist an area of weakness the stress will surely find it out.

The origin of the Eurasian chains has been ascribed to thrusts proceeding from the south and south-east. In this matter we must consider the tectonics of the world at large. We recall that the existing mountain ranges are nearly all of recent and, geologically speaking, of simultaneous origin. The American mountains have been moulded by west and east stresses; the Eurasian by

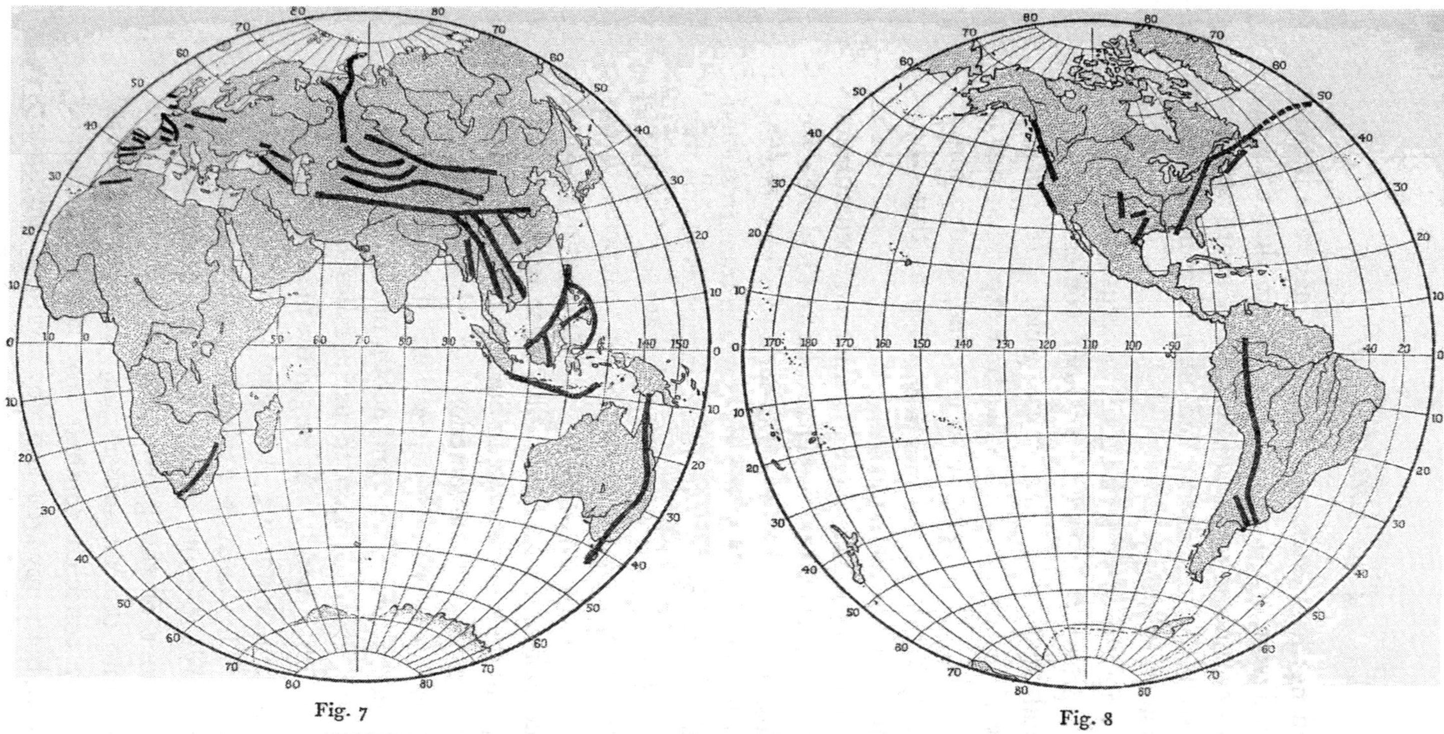

Fig. 7

Fig. 8

STEREOGRAPHIC MAPS OF EASTERN AND WESTERN HEMISPHERES

Showing ranges formed towards close of the Palaeozoic

south and north stresses. In short, we may take it that the east and west stresses were relieved by the heightening of the American ranges ; the north and south stresses were not so relieved, but expended themselves in elevating the Eurasian chains.

Attending the down-sinking of the crust, as we have seen, the pressure of the ocean floor comes upon the continental margin. This must yield. It thickens vertically, rising up and at the same time expanding downwards. Obedient to isostatic conditions, what is forced downwards is some eight times as deep as that which forms the raised borderland.

Then follows the long period of thermal accumulation in the substratum. The compensations first feel the effect of the loss of density. The borderland sinks, bearing downwards the adjoining crust over which transgressional waters flow. The geosynclinal sea originates in this way. It endures throughout the succeeding ages. The rivers that formerly carried their burthen of sediments to the ocean are now deflected into it. The horizontal pressure of the ocean floor may have, during these ages, ceased to operate; but the ever-increasing burthen of the sediments and the advancing liquefaction of the substratum maintain the geosynclinal depression. It develops, not without fluctuations of area and depth, into a great inland sea perhaps over two or more thousand miles in length and many hundreds in width. At length the period of another revolution comes about ; the geosynclinal sea grows shallower. The ancient pressure from the ocean is renewed. What must happen ? The deformed and weakened floor of the geosynclinal sea, for long ages depressed into the hot substratum, yields to the stress. Its patiently collected contents are laterally deformed and forced both upwards and downwards; the underlying continental rocks being forced still deeper into the magma, giving rise to compensations which may be many miles in depth. It is in this way that weakened areas, elongated, parallel with the coasts, originate. The stratified structure, centrally bowed downwards, is laterally weak towards compressive forces, and the supporting floor is softened by excessive thermal conditions of age-long duration. Barrell, writing in 1918, says:

'Willis showed that the deep subsidence of the centre of the geosyncline gave an initial dip which determined the position of yielding under compression. Laboratory experiments brought out the weakness of the stratigraphic structure to resist horizontal compression.'

In this great lateral movement, generally of several miles extent, much of the folding and over-thrusting recognized in the mountains takes place. It slowly develops throughout the long ages which witness these events; and we notice that throughout the process of its development, isostasy controls the vertical movements and leaves the newly-formed mountains in isostatic equilibrium.

But the mountain genesis is not yet finished. The events of the revolution terminate in the complete reconsolidation of the magma. Great isostatic forces now act on the newly-formed compensations, and the whole crushed and folded mass above rises yet higher. This final vertical movement is a characteristic event attending the close of orogenesis. Sometimes it is renewed at a subsequent revolution, as in the case of the Appalachians, which—mainly developed in the Permo-Carboniferous revolution—received a vertical lift of 2,000 feet in the Laramide revolution of Cretaceous times.

The mountain range has now come into being. And just as it has been born of the sediments washed from the raised borderland, so in turn its debris will weigh down the continental crust, and in due time give birth to the geosynclinal sediments of ages yet to come, and ultimately to a new generation of mountains, when the older ones are worn to shadows of their former greatness.

Dana long ago pointed out the striking fact that the greatest mountain ranges confront the widest oceans. The mechanical principles involved are simple. The wider Pacific represents a greater misfit than the narrower Atlantic. And the misfit has ultimately to be mainly accommodated within the yielding geosyncline. Hence a greater yielding of the western than of the eastern coasts of America was inevitable. Similarly the north and south misfit of much of the expanded surface crust of the earth finds accommodation in the complex folding of the Alps, Pyrenees, and the endless ranges of the Himalayas: the Eurasian chains ranging—as we know—approximately east and west.

The conclusions arising out of observational geology support the view that mountain-building is not in general a uniformly continuous process. There appear to be pauses followed by renewals of orogenesis. These renewals may occur after considerable time intervals. Study of the mountain range in many cases reveals such vicissitudes. To what causes are they due?

The most direct explanation arises out of probable, if not indeed inevitable, events taking place in a deep substratum. Its lower levels by reason of the very great pressure are under different physical conditions, and do not behave in the same manner as its upper levels. But although there may be no abrupt volume-change of the more highly compressed magma attending the accumulation of radioactive heat, it is manifest that a time must come when instability arises. A small upward displacement must then be attended with voluminal expansion, and the mass experiences an acceleration towards the upper parts of the substratum. Such vertical displacement is impossible during the period when the substratum is solid throughout. But when the upper parts become fluid and partake of tidal and convective movements, conditions arise promoting the upward displacement of deep-seated magma.

At intervals—possibly far apart in time—during the period of tidal movement there will be on this account recrudescence of energy beneath oceans and continents. Orogenic developments already in progress will pause ; there may be temporary recurrence of tensional conditions in the crust ; and then, attending the continued loss of heat, once more compressional effects and renewed orogenesis. In short, we find reason to expect just the very intermittence of events which has been so often referred to by the field geologist. Mountain-building would progress in successive stages governed by the cataclysmic escape of energy from the depths of the substratum.

This explanation of the phenomena of orogenesis appears to be not only in accord with their intimate character : it also accounts for the magnitude of the horizontal movements which geologists have inferred. We cannot indeed numerically evaluate the effects of the substratal displacements; but if the substratum extends to depths such as seismology indicates, obviously very great diastrophic effects must arise before equilibrium is restored throughout. The mountains are not built by one grand compressional effort, but by successive efforts of a substratum whose volume-changes fluctuate over long intervals of time. Even movements of pure and simple uplift, arising out of temporary congelation and affecting the deep compensation, may be supposed to intervene before the whole period of instability and of thermal discharge—which constitutes the revolution—dies down into quiescence, and the still longer waiting period of thermal accumulation sets in.

If we accept this view as best meeting the requirements of surface observations, and as the most probable outcome of the varying physical conditions which must prevail in the substratum, we must regard the orogenic results of a revolution as the net outcome of fluctuating losses and gains of heat: losses sometimes exceeding the thermal supplies to the upper layers of the substratum; orogenesis being at such periods renewed: this state being succeeded by renewal of tensional conditions attending accessions of energy from beneath; to be followed inevitably by fresh compressional forces applied to the continental margin.

There is, as might be anticipated, much difficulty in arriving at estimates of the amount of crust shortening or areal diminution represented in the formation of a mountain chain. Even when we deal with the most recently formed mountain ranges such estimates can only be very approximate. And, of course, in dealing with ranges due to older revolutions we encounter the fresh difficulties arising from the fact that denudation has swept the older elevations for the greater part away—or even completely away—as evidenced by the base-levelled remains of pre-Cambrian mountains or of the ancestral Rockies, or of the earlier Appalachians. Further, the history of most of the great ranges reveals successive orogeneses corresponding generally to the successive great revolutions. Between each of these some forty or more millions of years of denudative removal intervene. Witness the vast denudation of the Laramides at the

Two views of the Weisshorn (14,804 ft.)

The upper view (from the Mettelhorn, 11,188 ft.) shows the south face: the lower (from above Randa) shows the east side. Both ice and water are at work. The amount of denudation displayed by the Alps of Valais is the more impressive from the fact that these mountains are for the greater part built out of highly resisting rocks, hard gneisses, mica-schists and siliceous slates.

close of the Cretaceous and before the later Tertiary movements brought the present ranges into existence; or the almost incredible denudation of the Alps within recent times. Barrell relates how, in their explorations of the western mountains of America, Powell, Dutton, and Gilbert

'saw the stupendous work of denudation carried to completion again and again during the progress of geological time'.

In truth what we have left to our observation of the past features of the earth appears to be but a small remnant. Any reliable estimate of long-past orogenesis is, evidently, for ever beyond power of estimation.

A rough estimate, based on existing mountain ranges, of the areal surface reduction involved in orogenesis throughout geological time has been made by Jeffreys.[1] As some ancient ranges—e. g. the Appalachians—are necessarily included, this is probably an excessive estimate as applied to the surface reduction involved in the recent great Alpine Revolution of Tertiary times. His estimate comes out as 1,872,000 square kilometres.

Now if we suppose a depth of the substratum of 100 miles (160 km.) to take part in a great revolution, and that the sum of the fluctuating changes of volume of the one sign (either expansional or contractional) amount to 7 per cent. of the initial volume (see Chapter III), then the corresponding sum of the changes in the earth's radius would be about 6·8 miles. The surface changes of one sign would add up to 650,000 square miles or 1,700,000 square kilometres. This result would, evidently, be adequate to meet the gross surface contraction as estimated by Jeffreys.

As observed above, it appears to be more in accordance

[1] *Phil. Mag.*, December 1916.

with the phenomena of orogenesis to assume that this result is reached by fluctuating thermal gains and losses of heat attending accessions of energy from beneath.

We may deal with calculations respecting the crustal movements required for mountain-building in another and physically more comprehensible way.

Estimates of the shortening in a west to east direction of the earth's crust in the genesis of the Rockies—mainly but not altogether in Cretaceous time—have afforded 29 miles. There was no notable horizontal orogenesis in eastern North America at this time. Let us assume the areal reduction of the earth's crust to be resolved into east and west and north and south movements, the former going mainly to elevate the Rockies. Now the surface increase of 650,000 square miles arrived at above corresponds to an increase of the terrestrial circumference of 41 miles. Thus after allowing for the genesis of the Rockies we appear to have available a surplus for other and lesser east and west deformations.

The disposition of continental materials upon the globe is such that we have to look to Eurasia for any great seaboard ranging east and west along which we may seek for orogenic developments due to forces directed in north and south. Hence mountain ranges complementary to the Cordilleras of the Americas are represented by the west-to-east ranges of Eurasia. It is remarkable that the effects of the Laramide revolution referable to lateral crustal movements are not so conspicuous in Eurasia as in the Americas. Possibly the records are in part lost as having mainly affected the ocean floor.[1]

In the case of some of the revolutions, if estimates are correct, greater amounts of crustal shortening have to be accounted for in order to explain the folding of the

[1] See Chap. VIII, p. 135, for another suggestion.

mountains. Thus the total shortening in the case of the Alps has been estimated at 76 miles. This is mainly to be ascribed to the most recent world-revolution in late Tertiary times, known as the Alpine Revolution. The Himalayas were elevated during the same revolution, but the crustal shortening involved in the genesis of these mountains should not be added to that ascribed to the Alps, for these ranges extend over very different meridians and would absorb parallel crustal movements mainly directed north and south. Jeffreys quotes Oldham as estimating for this movement 100 km. (62 miles, about). These compressional movements appear to be the greatest hitherto claimed for the building of a mountain system with the exception of the Appalachians, for the elevation of which Keith has recently claimed 200 miles.[1] He considers, however, that the folding of these mountains was mainly due to the injection from beneath of vast masses of intrusive rocks. These are known as batholiths, and are a common feature of interior mountain structure; and in some ranges—for instance the Cordilleras of Western North America—are present upon an enormous scale.

These adjuncts of mountain structure—the batholiths—merit special consideration. Batholiths consist of masses of granite or anorthosite ranged in the direction of the mountain axis and evidently brought upwards from great depths by the instrumentality of the same forces as are responsible for the orographic developments. They undoubtedly originate within the continental crust. Daly and other petrologists regard them as the upper parts of abyssal dikes of enormous dimensions. The

[1] 'Outlines of Appalachian Structure,' *Bull. Geol. Soc. of America*, vol. xxxiv, pt. 2, 1923.

granitic mass which forms the core of the Leinster mountains of South-Eastern Ireland, and which was evidently forced into its position beneath the pre-Devonian slates by deep-seated pressure, is an example of a batholith on a relatively petty scale. It is probably about 100 miles in length. Anorthosite batholiths are mainly found among Archaean intrusives, along with gigantic granitic batholiths, and associated with vast fissure eruptions of basaltic magma.

Now it has already been stated that the closing phases of mountain elevation are known to be generally attended by great vertical movements; and a simple and natural explanation of these movements was deduced from the events which occur upon the final consolidation of the substratum around the compensations which support the mountains and dip deeply into the magma beneath the continental crust. Such compensations consist necessarily of continental rocks which have for long ages been at the full temperature of the magma and, at periods just antecedent to the advent of tidal movements, may be even above this temperature. Internally their proper radioactivity can be shown to lead to temperatures at which they would in some cases be expected to soften, if not even to liquefy. For instance (as we saw in Chapter III), within such a compensation as must support the Tibetan plateau, the average height of which above the sea is stated to be 15,000 feet, it may be shown that a central temperature of 1,500° C. may be expected to prevail. Hence a much softened core, chemically perhaps granitic in character, may exist. In the great vertical pressures from beneath, which act upon such a protuberant mass, the mechanism which brings about the intrusion of plutonic continental magmas into the heart of the mountain range finds direct and simple explanation.

The history of the batholith, as outlined above, agrees with Daly's statement that

'without known exceptions each batholithic invasion has followed more or less closely a period of strong crustal deformation affecting the older formations of the same region '.[1]

Consider this statement of Barrell's :

'Many others have found that girdling the world a large part of the mountainous relief is due to vertical elevatory forces acting over regions of previous folding and overthrust. In addition great plateau areas of unfolded rocks have been bodily lifted one or two miles, or more, above their earlier levels. They may be broad geanticlinal arches or bounded by the walls of profound fractures. The linear mountain systems made from deep troughs of sediments have come then to be recognized as but one of several classes of mountains.'

Now vertical movements have already been referred to as necessarily acting over regions of previous folding and overthrust. For this folding and overthrusting involves the genesis of the greater compensations, and in due time, when the revolution which brings about the folding has run its course, the succeeding period of congelation occasions the great vertical thrusts which complete orogenesis. Their importance has already been dwelt on. Barrell recognizes them as constituting a distinct orogenic feature. For in fact the great plateau areas also have their compensations and, especially if fractured by tensile stresses preceding mountain-building, must arise and stand 'bounded by the walls of profound fractures'.

Some high authorities regard these great vertical movements attending mountain elevation as largely responsible for much of the visible crustal deformation, shearing and folding of the rocks and similar orogenic phenomena. Again Marcel Bertrand, and others following him, have recognized in the mere settlement of the sediments in a

[1] *Igneous Rocks*, p. 86.

convergent geosyncline a necessary source of lateral compression and folding taking place while the sediments are far beneath the surface of the continent. Both views are probably true.

At the present time it is doubtful if any of the inland waters of the earth can be regarded as occupying geosynclines. It is a period of high land elevation, and transgressional seas do not now exist. The great rivers which formerly fed the geosynclines have also vanished, and new ones have taken their place which traverse often great continental distances before they discharge their debris into the ocean. Witness the Mississippi, and consider that the enormous burthen of sediments upon which New Orleans is built, and which has carried the land many hundreds of miles into the Gulf of Mexico, in former ages would have been retained in some great geosynclinal basin. Therein it would collect, building up beds of sedimentary rocks out of which mountain ranges would ultimately be formed.

The great vertical depths sometimes claimed for accumulations of mountain-building deposits have, however, been questioned. And, indeed, the limitations imposed by isostasy and the probable depths of the transgressional seas must be taken into account before they can be accepted. But this fact is not opposed to such estimates as those advanced by Chamberlin and Salisbury [1] that the height of the mountains developed in the Laramide Range in late Cretaceous time was 20,000 feet, and that, owing to the further elevation which has since taken place, from 32,000 to 35,000 feet would be their present height if erosion had not undone its own work and hurried some twenty thousand feet of the range into the rivers.

[1] *Geology*, 1906, iii. 163.

VIEW FROM MONTE PIANO, looking south and embracing the Marmarole Mts. (10,266 ft.): Antalao (10,680 ft.) and Sorapis (10,860 ft.). The deeply cut valleys responsible for this magnificent scene are the work of frost and thaw arising out of diurnal and annual changes of temperature acting upon vast deposits of bedded calcareous rocks.

The crustal movements we have been considering are *recurrent in character and take place in the same geosynclinal area.* Schuchert has defined the geosynclines of North America, showing that in early Palaeozoic times they were already in existence : a fact which is probably true of all the great geosynclines of the earth. Their history has been very various. The oldest of the North American geosynclines, the Cordilleran, accumulated sediments till Cretaceous times; the successive deposits of Proterozoic and Palaeozoic age show conformable relations throughout; and this orderly sequence of deposition continued to the close of the Jurassic. The first Rocky Mountains geanticline came into existence in closing Carboniferous (early Pennsylvanian) time. The whole of this high anticline was, however, removed by denudation by late Jurassic time. In the Laramide (closing Cretaceous) the Rocky Mountains (of our day) were elevated in this ancient Palaeozoic geosyncline, and were again further uplifted, through several thousand feet, in the Pliocene. And, indeed—

'especially during the Miocene, and less in the Pliocene, the entire area of the overlaps of the Pacific in North America was being elevated, folded, faulted and thrust eastwards into the Pacific system of Mountains'.[1]

The Appalachian trough existed far back in the Proterozoic. The sediments, poured into it throughout the succeeding ages, came from the east and south-east; for an easterly extension of North America existed at that time which has since become submerged in the Atlantic. The supply was, however, by no means uniform. The land was changing in level and here

'we again see what is now so well known, namely, that the trans-

[1] Schuchert, *Geology*. See also *Bull. Geol. Soc. America*, vol. xxxiv, part 2, 1923, for a fine exposition by the same author of the history of North American Geosynclines.

gressions of the ocean upon the continents are periodic in appearance, and more or less irregular in their spreadings'.

The succeeding history of this great range is extraordinary, and the facts set forth by Schuchert (loc. cit.) most instructive.

The movements of the Appalachian Revolution brought up, folded, overthrust, and often metamorphosed, the age-long accumulations that had been forming since early Proterozoic times. In the Permian, after a loss by denudation amounting to several miles in depth, the already worn mountains were vertically re-elevated some thousands of feet.

The recurrent character of orogenesis is similarly revealed in the Alpine system of Europe. Three main periods of disturbance have been assigned: closing Palaeozoic (Appalachian Revolution ?), closing Cretaceous (Laramide Revolution), and closing Tertiary (Alpine or Cascadian Revolution). And so also we might find in other great ranges revelations of the same sort: the Pyrenees, the Caucasus, &c.

Much remains unknown respecting the Andes. An ancient geosyncline, about 5,000 miles long, separated from the Pacific by a raised borderland, appears to have persisted till Cretaceous times; frequent uplifts affecting the borderland. In the Laramide Revolution the accumulated sediments were folded, thrust towards the east, and uplifted. Great denudation reduced the height of these ranges, but they were again raised vertically in Tertiary times, and, finally, after the Alpine Revolution, in Pleistocene times. Douglas, writing of the Andes of Peru and Bolivia, considers that the mid-Tertiary uplift amounted to at least 14,000 feet, and that great outbursts of volcanic activity accompanied the movement. He concludes that these mountains are mainly the work of vertical movements, of which he records five. These appear to succeed the Caledonian, the Appalachian, the Laramide, and the Alpine Revolutions. An uplift also occurred in post-Jurassic times attended by batholithic invasion. This movement seems to coincide chronologically with the movement which affected the Sierra Nevada, the Coast Range, and the Humboldt Range of North America in the same manner.[1]

The explanation of mountain genesis which arises out of a periodic change in the density of the substratum is distinguished from other suggested explanations of orogenesis in the fact that it provides at once cause for lateral as well as for vertical movements, and the order in which these take effect is consistent with recorded observations.

The igneous activity very often associated with mountain-building must be regarded as testimony to the depth in the continental crust to which the orogenic movements extend. The fact that the erupted materials are most generally basaltic or andesitic in character is further evidence that these movements reach downwards into the substratum itself.

There is hardly a great mountain chain on the globe that is not associated with contemporary volcanism. Thus the elevation of the Alps was attended by volcanism in many surrounding regions. In Auvergne, where from Miocene to Pliocene times basaltic flows from vents and craters (commencing with trachytes) flooded the surrounding country: in the Eifel, where craters formed in Miocene time and active till the Pleistocene, poured out mainly basaltic lavas: in North Italy, where volcanoes also broke out. Similarly, the resurrection of the Pyrenees

[1] *Q.J.G.S.*, 56, 1920, p. 1.

was associated with Eocene volcanoes, which poured out basalts in Catalonia. The chief summits of the Caucasus, like those of the Andes, are themselves volcanoes. Many other cases might be cited.

Lindgren, referring to the volcanism of the Pacific coast of North America, describes how the era of basaltic effusions began in the Eocene and reached its maximum in the Miocene. Enormous volumes of lava were poured out in the north-western areas of the United States.

'The fundamental fact in the Cordilleran region is that the igneous activity began along the present Pacific coast-line and gradually extended eastward. The initial stresses, primarily causing the rise of the magma and secondarily the deformation of the sediments, therefore came from the Pacific side. They were abyssal, extending to depths of many miles.' 'Everywhere intrusion (of batholiths) corresponds to uplifting, and the evidence, it seems to me, is entirely favourable to simultaneous uplift and intrusion.'

Although andesites in some areas predominate,

'on the whole basalt is probably the most widespread of the Tertiary effusives'.[1]

The whole of the Pacific is, in like manner, girt round with volcanism ; for, in truth, along every part of its shores orogenesis and its attendant deep crustal movements have been at work. They extend from the Andes to the Antarctic and confer on Eastern Australia its mountainous border, of which, possibly, New Zealand once formed a part. The feebler orogenesis of the Atlantic has left no such records behind it, and little coastal volcanicity has affected it save in its northern regions, where the Icelandic volcanism and the Hebridean plateau basalts form a basaltic boundary to its waters.

If we study the oro-bathygraphical chart of the world, given at the conclusion of this volume, we perceive how

[1] 'The Igneous Geology of the Cordilleras and its Problems,' *Problems of American Geology*, 1915, pp. 273 et seq.

the sea and the volcano appear generally associated. This fact has given rise to the view that some causative connexion must exist between them. But, in truth, volcanoes can and do occur far from the ocean. What then is the true nature of the association? Just this: The sea-floor has been the passive instrument of orogenesis. It has forced inwards the continental margin and deformed and rent the continental crust to its depths. The lava wells out, caught in gaping fissures beneath and forced upwards. And as the stresses increase, the sediments themselves, folded and crushed, emerge above the surface of the geosynclinal sea. The batholithic masses of continental rocks, possibly not at a very high temperature, but softened by an aqueo-igneous action, arise along with the sediments to form the heart of the youthful ranges.

But the ocean waters take no direct part in the volcanism attending these movements. Their presence has no causative significance.

It is certain that very much remains to be learnt about the nature of geosynclines. In the case of the Mediterranean (in former times expanded into a far larger inter-continental sea which geologists call the Tethys), the widely extended shallows along the shores of the Tethys appear to have received the sediments, and there they collected age after age, sinking ever deeper till the period for the genesis had arrived.

From the ancient Tethys the Himalayas arose after a period of deposition longer than any other known: beginning far back in pre-Cambrian times and with certain intervals of local uplift, persisting till the Miocene. Comparatively recently, therefore, they have uplifted their burthen of sediments collected throughout the whole of geological time, and to-day maintain them, clothed in their covering of snow, higher in the heavens than any other mountain summits of the globe. But we know from our studies of the past that they will reign for a comparatively brief period. The vast sedimentary collections of the Indo-Gangetic plain and of the deltas of the Ganges are eloquent of the fleeting existence of even the mightiest mountains of the earth.

## APPENDIX TO CHAPTER VII

### *The Cyclical Changes in the Substratum*

The cycle of changes progressing in the substratum and overlying crust which result in transgressional seas and mountain elevation are comparable with those of a heat engine, possessing source and refrigerator and a working substance; the latter experiencing a volume-change attending change of state, and in that way being able to do external work.

We may describe the cycle as follows: We start with the working substance (the basaltic substratum) in the process of accumulating radioactive heat. The resultant events will not progress exactly at the same rate at different depths in the substratum. We fix our attention upon a certain mass which is some distance down and which is accumulating heat from its own radioactivity, i.e. the 'source'. In the diagram we suppose its state as regards pressure and volume to be defined by the position of the point *A*. As it accumulates heat it gradually melts, increasing in volume along the isotherm *A*—*B*. At *B* it is completely melted, and in virtue of its reduced density it floats upwards, for it will be somewhat superheated respecting the overlying magma. Its pressure falls as it moves upwards till its pressure is that of *C*. It now begins to give up heat to the ocean (i. e. the 'refrigerator') and to shrink in volume along the isotherm *C*—*D*. At *D* it is completely solidified, and because of its increased density it sinks downwards, its pressure augmenting to that of point *A*. The cyclical changes now start afresh.

As regards the geological events arising out of these changes, we may describe *A*—*B* as representing the long period of increasing transgressional seas. The description of *B*—*C* occupies a relatively short period during which transgressional seas are at their height. From *C* to *D* is the mountain-building period. Heat is escaping from the magma; its volume is diminishing and the crushing of the continental margins by the down-sinking ocean floor is increasing. The sinking of the congealed magma along *D—A* occupies but a relatively short period. The area *ABCD* represents the work accomplished attending the sequence of changes, and may be regarded as mainly expended in orogenesis.

Thermo-dynamic cooling due to fall of pressure along *B—C*, and equivalent heating due to rise of pressure along *D—A*, are not considered, as probably having but little influence on the nature of the events.

We notice that the source and the working substance are enclosed beneath a lagging some 30 km. in thickness and most cunningly contrived so as to conserve the energy within. The insulation is secured by the lagging itself being raised in temperature by its own radioactive heat.

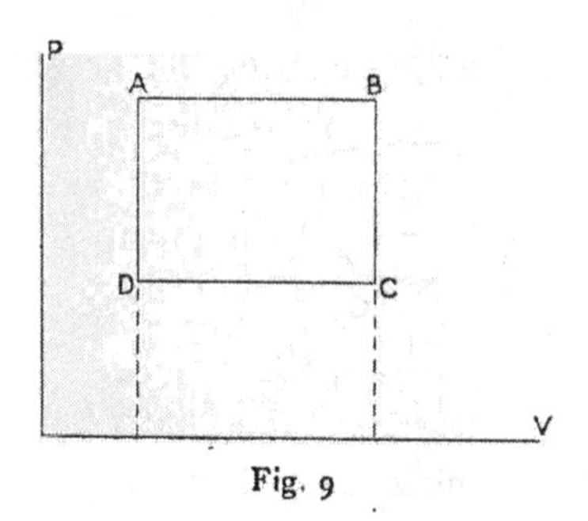


Fig. 9

Not the least beautiful part of the whole mechanism is the part played by our one and only satellite. The perpetual supply from the 'source' must certainly end in disaster to the world above, if it was not for lunar interference. And this interference automatically comes into operation exactly at the critical moment. For so long as the substratum is solid the moon is powerless—even with solar aid—to raise tides in the substratum and to shift the outer crust of the earth. But just when the substratum melts and further thermal supplies must involve rising temperature and, ultimately, melting continents, astronomic interference comes in. The ocean floor is caused to thin out under the fiery tides raised in the underworld; the ocean rapidly absorbs the heat and all danger is averted. The shifting of the earth's crust over the substratum brings the same redemption to the whole earth.

## VIII

## THE REVOLUTIONS

In Chapters VI and VII we reviewed the conditions which control the birth, advent, consummation, and decay of world-revolutions. These revolutions may vary in magnitude. Not only our reasonable expectations, but the geological evidence, point to this conclusion.

We anticipate variability, because a period of time sufficient to liquefy the upper layers of the substratum will not suffice to destroy the equilibrium of the depths. On the other hand, if, in the depths, the accumulation of radioactive heat has progressed for a very long period, the liquefaction of the upper layer and the attendant tidal disturbance must result in precipitating the break-up of the deeper layers and a far more diastrophic world-revolution must result.

The criterion of a revolution appears to be the advent of tidal movements of the crust; for these movements control the escape of heat and therefore complete the cycle. The conditions for tidal intervention involve a break-down of the magma extending for some few miles below the continental base. However, the change of physical properties downwards must be very gradual, which leads us to the view that even the most superficial revolution must involve considerable depths of disturbance.

The magnitude of the orogenesis attending a revolution is best expressed in terms of the circumferential variation in the size of the globe. For, as we have already seen, it is this which defines the range of the horizontal forces developing out of the reaction between the ocean floor and the continents.

On the assumption that 100 miles in depth take part in the genesis of a revolution, and that there is an integral volume-change of 7 per cent., the circumferential increase is about 40 miles, and as a sufficient approximation we may assume that the increase will be proportional to the depth which takes part in the revolution. Now it might well be that a world-revolution might arise in which only half this circumferential change was involved. And again we might expect that others might arise which were attended with circulation of the substratum to depths considerably greater than 100 miles.

But there are other possibilities—or, rather, we should say probabilities. A revolution after it had run a certain course and was perhaps in process of decline might be reinforced by fresh supplies of heat from beneath. Obviously this might very naturally arise if the substratum is as deep as seismic evidence would appear to indicate. The effects at the surface must be a renewal of transgressional seas already waning; and, possibly, a renewal of orogenesis. So long as tidal movements are in progress such reinforcements are at any time probable. The long waiting interval of accumulating latent heat only properly begins when tidal movements have finally ceased. Clockwork regularity of the great phenomena of the cycles could only be expected upon conditions which would be ideal and not real. In point of fact, the history which we have now to review as best we may points to just such irregularities as might be anticipated.

If our knowledge of the physical conditions controlling revolutions is limited, our knowledge of earth-history is also limited. For in truth we have only the deeply weathered surface and almost obliterated history of a world presented to our investigation. When, therefore, we approach the subject of earth-history in relation to isostasy and radioactivity, we find that we have but an imperfect record upon which to rely in trying to decipher the revolutionary history of the earth.

The Archaean—as Chamberlin and Salisbury point out—is a universal geological system in the sense that Archaean rocks appear to underlie all other surface deposits of the earth. But in Canada and the Northern States of America they extend over vast areas as surface materials, and hence their study has been prosecuted in these regions more successfully than elsewhere. Still much remains in doubt. The difficulties in the way of the correlation of azoic rock-systems in areas so vast as those dealt with in North America are great indeed. To these difficulties confusion in nomenclature has added not a little. Again the derivation of the highly altered materials, whether from an igneous or from a sedimentary source, is often so obscure that the interpretations of even skilled observers have occasionally been contradictory. Planes of unconformity used as planes of reference have led to divergent views as to their significance. Their identification has often been matter of dispute. In recent years these sources of confusion have been considerably reduced.

A study of the comparative table submitted by Coleman to the Geological Congress of 1913 shows that two great sedimentary series are fundamental: the Grenville Series, said to be some 94,000 feet thick; and the Sudbury, not less than 20,000 feet. These great series do not probably overlap in time. Both these, together with a series mainly of basaltic eruptives (the Keewatin), were finally intruded by vast uprisings of granites (largely batholithic) to which the historic name of Laurentian is assigned. The attendant igneous phenomena are admitted by all authorities as far transcending any known subsequent phenomena of the kind.

On this and much other evidence it is concluded that the Laurentian was a great orogenic and diastrophic period; a world-revolution upon an enormous scale. The mountains themselves have long ago been base-levelled by denudation to the existing peneplanes of the Canadian Shield, where they extend over an area of at least a million square miles.

In these phenomena we are presented with a great primary revolution (the Laurentian Revolution) in which vast precedent sedimentation was overwhelmed by uprising granitic eruptions and basic lava-flows.

Calling the sedimentary series so overwhelmed 'The Archaean', Chamberlin and Salisbury say—

'Not unlikely it consists of a much more extended succession of formations reaching indefinitely downwards'.[1]

Many authorities would recognize in pre-Cambrian times not one but three great revolutions as having occurred. The first (the Laurentian) succeeding the formation of the Grenville deposits, the second (the Algoman) the formation of the Sudbury deposits, the third (the Killarney or Grand Canyon) closing pre-Cambrian time. The first revolution is beyond doubt far the more important. The matter is too complex for discussion in this place.

Schuchert and Barrell, who have given much study to this great subject, signalize six major world-revolutions as occurring in Geological time, as follows:

1. The *Laurentian* coming after the deposition of the Grenville series of sediments and the Keewatin eruptives, &c.

2. The *Algoman* coming after the deposition of the Sudburian sediments and closing Archaean time.

3. The *Grand Canyon* or *Killarney* coming after the deposition of the Huronian and Keweenawan sediments and closing pre-Cambrian time.

4. The *Appalachian* closing Palaeozoic time.

[1] *Geology*, p. 142.

5. The *Laramide* closing Mesozoic time.

6. The *Cascadian* closing Tertiary time and extending into Quaternary time.

In this classification the revolution is the period of mountain-building or orogenesis. As we saw in the last chapter, this period is led up to by a very long era of slowly increasing and fluctuating transgressional seas, characterized by sedimentary deposition in the geosynclines, and finally of retreat of these seas culminating in the period of mountain-building.

Sonder in a recent study of the phenomena attending and leading up to a revolution recognizes—(1) a long continental stage; (2) a period of submergence by oceanic transgressions; (3) one of fluctuating levels; and (4) one of emergence and mountain-building. His recognition of a fluctuating (*Wechsel*) stage is of interest, although fluctuations attending the submergence of the continents have long been recognized. Applying these views to geological history he allocates the events of the cycles as follows:

| | |
|---|---|
| Cainozoic | *Emergence and Mountain-building.* |
| Cretaceous | Maximum transgression and fluctuations. |
| Jurassic | Submergence. |
| Triassic | Continental period. |
| Upper Carb. and Permian | *Emergence and Mountain-building.* |
| Devonian | Submergence. |
| Lower O.R.S. | Continental period. |
| Late Silurian | *Emergence and Mountain-building.* |
| Cambrian | Submergence. |
| Eo-Cambrian | Continental period.[1] |

Marr,[2] writing at an earlier date and discussing the geological history of the British area, recognizes—(1) a

[1] Sonder: 'Die erdgeschichtlichen Diastrophismen im Lichte der Kontraktionslehre,' *Geol. Rundschau*, xiii, 1922. [2] *Geology*, p. 149.

*continental* (i. e. a high-level and orogenic) *period* in the pre-Cambrian followed by a 'marine' (submergence) period in Lower Palaeozoic times; (2) a *continental period* at end of Lower Palaeozoic and a marine period in Carboniferous times; (3) a *continental period* in Permo-Triassic times and a marine period in Jurassic to early part of Tertiary; (4) a *continental period* culminating in Miocene times.

We may tabulate and compare these views, adding the revolutions as recognized by Haug,[1] and the historical sketch given by de Lapparent.[2]

| | *Barrell* and *Schuchert* (1924) | *Haug* (1921) | *Marr* (1898) | *Sonder* (1924) | *De Lapparent* (1900) |
|---|---|---|---|---|---|
| Recent | | | | | |
| Pleistocene | | | | | |
| Pliocene | | | | | |
| Miocene | Cascadian Revolution | Alpine and Dinaric | 'Continental' | 'Emergence' | Great Alpine Uplift. |
| Oligocene | | | | | |
| Eocene | | | | | Uplift of Pyrenees and Apennines. |
| Cretaceous | Laramide Revolution | | | | |
| Jurassic | | | | | |
| Triassic | | | | | |
| Permian | Appalachian Revolution | Hercynian | 'Continental' | 'Emergence' | Hercynian Chains formed. |
| Carboniferous | | | | | |
| Devonian | | | | | |
| Silurian | Caledonian disturbances | Caledonian | 'Continental' | 'Emergence' | Caledonian Chains formed. |
| Ordovician | | | | | |
| Cambrian | | | | | |
| Keweenawan, Huronian | Killarney Revolution | Huronian | 'Continental' | 'Emergence' (?) | Huronian Chains formed. |
| Timiskamian | Algoman Revolution | | | | |
| Loganian | Laurentian Revolution | | | | Continents outlined. |

[1] *Traité de Géologie*, 1921, vol. i, pp. 527 et seq.
[2] Ibid., 4th ed., 1900, iii, p. 1864.

There is, on the whole, very general agreement between the historical schemes outlined by these distinguished authorities. The variations arise out of differences in the degree of diastrophism in the two continents. Thus the Caledonian was more accentuated in Europe than in North America; whereas the Laramide was more accentuated in America than in Europe.

The latter case is of special interest. In America, both North and South, there were conspicuous orogenic developments in closing Cretaceous times. The great deposition of Palaeozoic and Mesozoic sediments over the site of the future Laramide Mountains had come to an end. In closing Cretaceous times, thrusting from the west folded these sediments into mountains. A height believed to have been 20,000 feet was ultimately attained, the horizontal compressional movement being about 25 miles.

'The close of the Laramide was, *par excellence*, the period of orogenic movement in the western part of North America. . . . The folding movements extended from Alaska in the north to Cape Horn in the south, more than a quarter of the circumference of the earth.' [1]

The transgressional seas retreated off the American continent during closing Cretaceous times.

This great orogenesis is positive evidence of a world-revolution. Some European geologists, finding no similar orogenic developments in Eurasia, would merge this time-interval into the preparatory period of the succeeding revolution: that which had its orogenic climax in Miocene and Pliocene times. But once it is recognized that it is the movement of the sea-floor which is mainly concerned in orogenesis, we have to admit the possibility that the diastrophism may have been expended upon a region invisible to us and have taken the form of deformation of the ocean floor.

[1] Chamberlin and Salisbury, *Geology*.

There is, in this case, another alternative. There is strong evidence, both palaeontological and structural, for the former existence of a great continent extended equatorially from South America to Africa and from Africa to India, occupying the northern part of the Indian Ocean and reaching to Australia. As will be shown in Chapter X, a continental tract so extended in longitude (about 230°) would be of doubtful stability.

It was at or near the close of the Cretaceous and dawn of the Eocene that this great continent (named by Suess Gondwanaland) broke up and, as a continent, disappeared.[1] This event was, so far as we know, synchronous with the overwhelming of Peninsular India by a mile-deep covering of the basaltic substratum. And, further, it is believed that the diastrophism in Gondwanaland was connected with the movements which led to the outflow of the Deccan; for the fissures which emitted the lava had the same east to west trend as ancient faults which traversed the northern part of Gondwanaland and along which the later movements were directed. Madagascar, the Seychelles and the Falkland Islands are by many regarded as fragments of the dismembered continental tract. 'The breaking up of Gondwanaland cannot be put later than the end of the Cretaceous or early Tertiary.'[2]

It is possible that the diastrophism attending the close of the Cretaceous was concerned in its disappearance. Upon such a basis, purely hypothetical it is true, the meridional expansion of the ocean floor might be accounted for, the latitudinal being accounted for in the orogenesis of the Western American ranges.

According to Barrell's and Schuchert's spacing of the revolutions, the interval between the Laramie (closing

[1] See Reed's *Geology of the British Empire*, p. 288, for an account of this matter.

[2] Reed, loc. cit.

Cretaceous) and the closing Oligocene must have been sufficient for the substratum to accumulate sufficient heat to bring about yet another period of fluidity. Or must we assume the intervention of super-heated magmas from beneath?

There is much evidence that the Eocene period was a long one. The biologic evidence for this has often been adduced as revealing a great break between the two systems. For in fact many divisions of organic life show remarkable developments in the period separating the latest Cretaceous from the early Eocene life, as we find them preserved in the deposits. Perhaps the most striking advances were in Mammalian evolution. Much purely geological evidence supports this view. The Laramie deposits are generally unconformable with the Eocene.

> 'The break between the Laramie and the Eocene is locally a great one—has even been regarded as one of the greatest breaks recorded in the strata of the continent.' [1]

In Colorado the erosion between the epoch of the Laramie proper and that of the Araphahoe formation above the unconformity is thought to have been very great. Cross estimates it to have been 14,000 feet. The time involved must, therefore, have been long. So also, locally, the Cretaceous suffered as much as 7,000 feet of erosion after the post-Laramie uplift and before early Eocene formations.[2] The maximum thickness of the Eocene near the 40th parallel has been estimated by King as 10,000 feet.

An estimate of the time-interval from Eocene to Oligocene by the Uranium-lead ratio (see Chapter IX on Geological Time) gives thirty-four million years, which is probably excessive.

[1] Chamberlin and Salisbury, loc. cit., iii. 155.
[2] Ibid., 158, 213.

With the exception of the revolution which ushered in our own times—and to which we shall presently refer—no one of the recorded revolutions stands out so clearly as a major historical event as that which filled much of Carboniferous and Permian time: the Appalachian.

Throughout the greater part of the Carboniferous fluctuating transgressions invaded the continents. In North America these began in early Mississippian time (Lower Carboniferous). Lesser deformations arose out of these movements, and the changing geographical conditions continued into Pennsylvanian time (Upper Carboniferous). This is the period of fluctuating levels which attends the liquefaction of the substratum and the prevalence of tidal movements under the influence of tide-generating forces.

In Central and Western Europe there were similar fluctuations, and in Europe, as also in North and South America, in the Upper Carboniferous and Permian, mountain-building forces were at work. In Europe the Palaeozoic Alps arose—great fold-mountains facing the immense ocean stretch which lay to the South. The worn-down remains of these mountain systems are to be seen in Germany, France, Belgium, England, and Ireland—the Armorican and Variscan Alps of Suess.

In North America orogenesis progressed both on the east and west sides of the continent. Successive movements were responsible for the creation of the Appalachian ranges on the eastern side. Upon the western side a successive series of elevations is revealed in marine and continental deposits. These movements extended over a long interval of Carboniferous and Permian time. They illustrate the essentially complex and fluctuating character of the changes progressing in the substratum.

We shall now pursue the development of events in the last great revolution preceding our own time as deciphered by geologists in America and Eurasia.

The events which close Cretaceous time tell us plainly of a substratum which has grown dense by reason of its congelation and is able vertically to displace the great compensations of the Laramide Mountains, of the Antilles, of the Andes, and of the ancient Appalachians. The vast seas which had overspread very much of America and Eurasia had disappeared. But in Middle Eocene they begin again to appear in Eurasia and in North and South America. We judge from their growth that the substratum is again losing density and the continents are sinking gradually deeper into the sustaining basaltic magma. But as yet there are no transgressions comparable with those of Middle Cretaceous times. Contrast the world-maps of de Lapparent for the periods of the Middle Chalk and of the Middle Eocene. But, plainly, the advent of a great revolution is affecting the continents of the earth.

By the middle of the Eocene the Tethys extends to the Pacific, and commencing transgressions invade the western borders of North America, submerge the Antilles, and enlarge the Gulf of Mexico. The Hebridean basaltic floods pour out over the Atlantic floor and attain some thousands of feet in thickness.

During Oligocene time loss of density continues. North America loses about 10 per cent. of its area in border transgressions. The Tethys swells to its greatest dimensions. The liquefaction of the substratum appears to have attained, or to be near, its maximum. The crustal tension is great. At this time the continent of Africa is rent asunder from south to north by irresistible forces. Great outpourings of basaltic lava occur in South-eastern

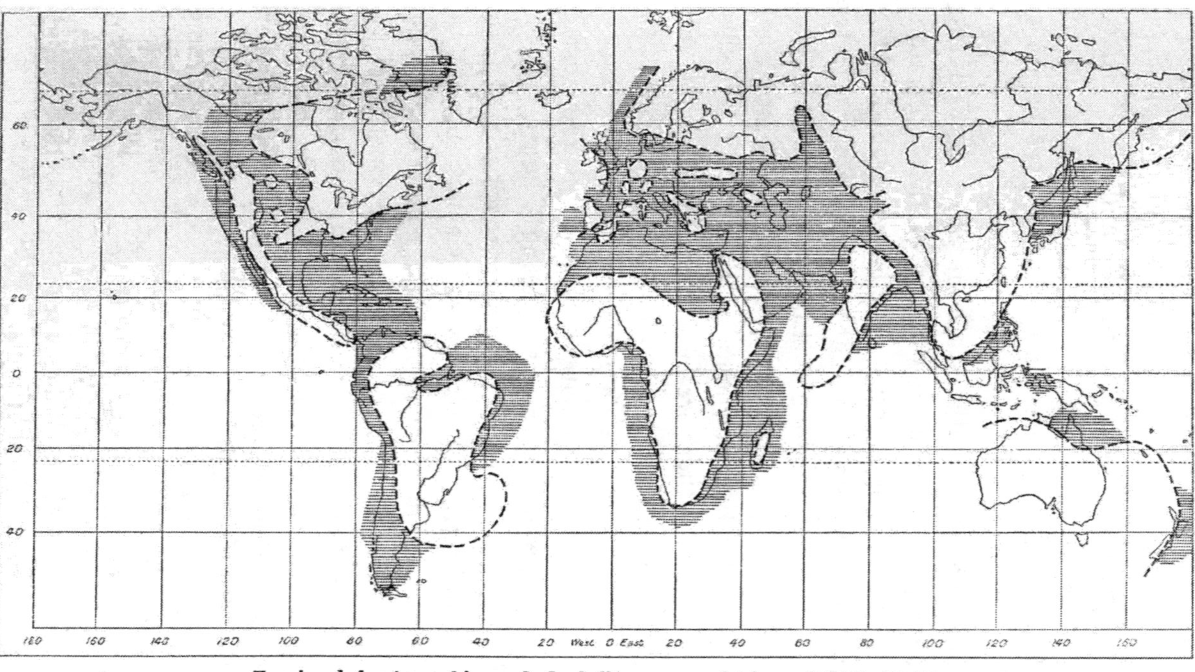


Esquisse de la géographie vers la fin de l'époque emschérienne (Middle Chalk)
After de Lapparent

and North-western Australia. The long rift valley of South Australia, mainly meridional in direction, as well as meridionally-directed fracture-lines along the eastern side of that continent, develop in middle and late Tertiary. And now commencing orogenesis is indicated in the Himalayan area, but the Tethys has not yet begun to shrink.

This period marks the close of those ages during which heat has been accumulating in the substratum. Tidal movements of the crust have long prevailed. Heat is passing into the ocean, and the world is preparing for the reversal of all the age-long events and movements which witnessed such great advances in organic evolution and such changes in the continental areas. At the very close of Oligocene time there are indications that the magma is again gaining in density; but mountain-building is not yet begun, or is in an incipient stage only.

There is much fluctuation of area in the surface waters of America and Eurasia during Oligocene times. The Tertiary deposits of Europe, more especially, reveal this phenomenon. Now from what has been already said we readily find an explanation of these movements. For these fluctuations tell us that tidal movements of the continents and ocean floor now prevail, and that vicissitudes in isostatic forces, arising out of density variations in the substratum traversed by the slowly shifting continents and oceans, affect the continental levels.

In the progress of events in the Miocene we find plainly revealed the effects of advancing thermal dissipation and the attendant increase of magmatic density. The mountains are rising. In early Miocene times the Eocene Nummulitic sediments are uplifted in the growing ranges of Eurasia and of India. Thrusts from south and south-east fold the Alpine system of Western and Central

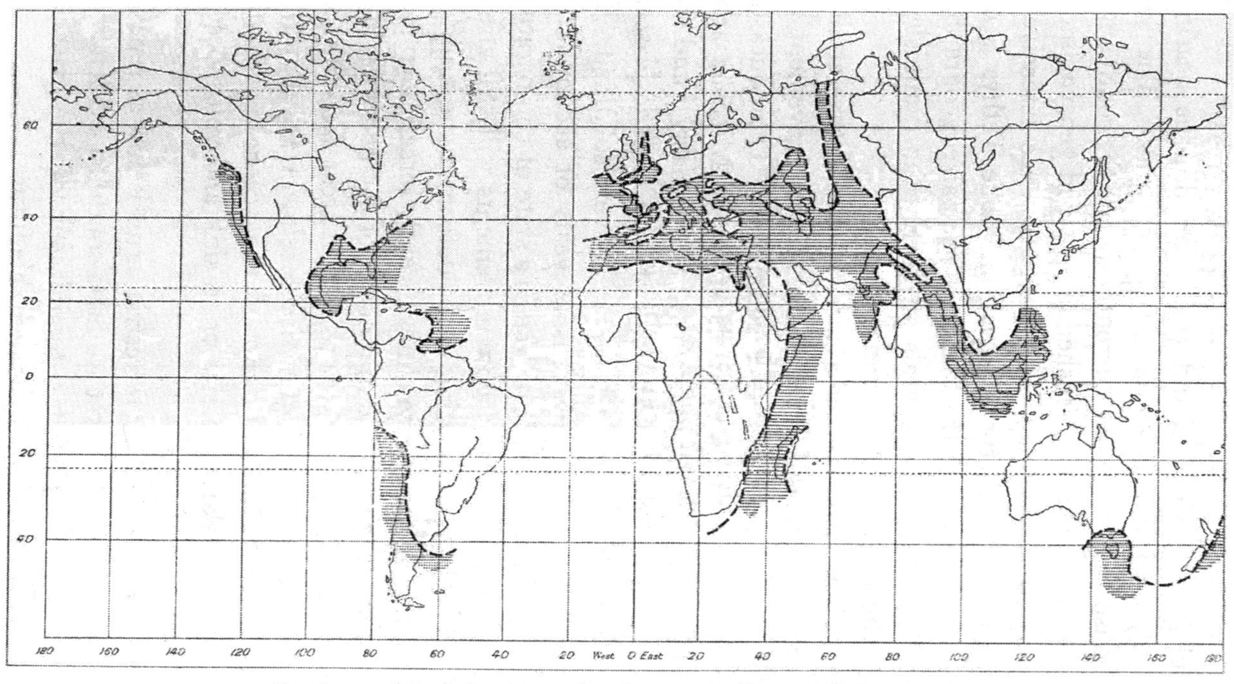


Esquisse partielle de la géographie à l'époque lutétienne (Middle Eocene)
After de Lapparent

Europe. The earth's crust is sinking downwards as the substratum shrinks, and the ocean floor is bearing irresistibly against the continental margins. The increased isostatic forces concurrently elevate the continents, and more especially the crushed and loaded geosynclines, relatively to the oceans. The Himalayas are extensively folded and gain greatly in elevation. The Tethys is no longer connected with the Pacific. The palaeogeographic map of de Lapparent reveals strikingly the diminished area of the ancient Mediterranean Sea.

In the next period—the Pliocene—the Tethys attains its present form and area. The vertical movement of the Western European Alps is completed. The Caucasus rise vertically bearing upwards strata of Miocene age. The third and greatest upheaval of the Himalayas and the vertical uplift of the Rockies and of the Colorado plateau take place. The substratum is effectively solidified. It has yielded up many millions of years of accumulated energy, and for a time it will be in a state of comparative rest. But for long ages slow movements will still affect the land. Energy lingering in the deformed and uplifted mountains; in the geosynclines recently forced deep into the magma (the invisible mountains of the underworld); in the flexured ocean floor; in the great laccoliths of still molten lava; will give rise to slow vertical movements of the continents and of the ocean floor; will cause oceanic islands to rise or to sink and volcanoes to eject their lavas.

It is probable that the terrestrial crust is now as reposeful as it will ever be; yet we have continual evidence of slow crustal movements—reminiscent of the past or prophetic of the far-off but inevitable future.

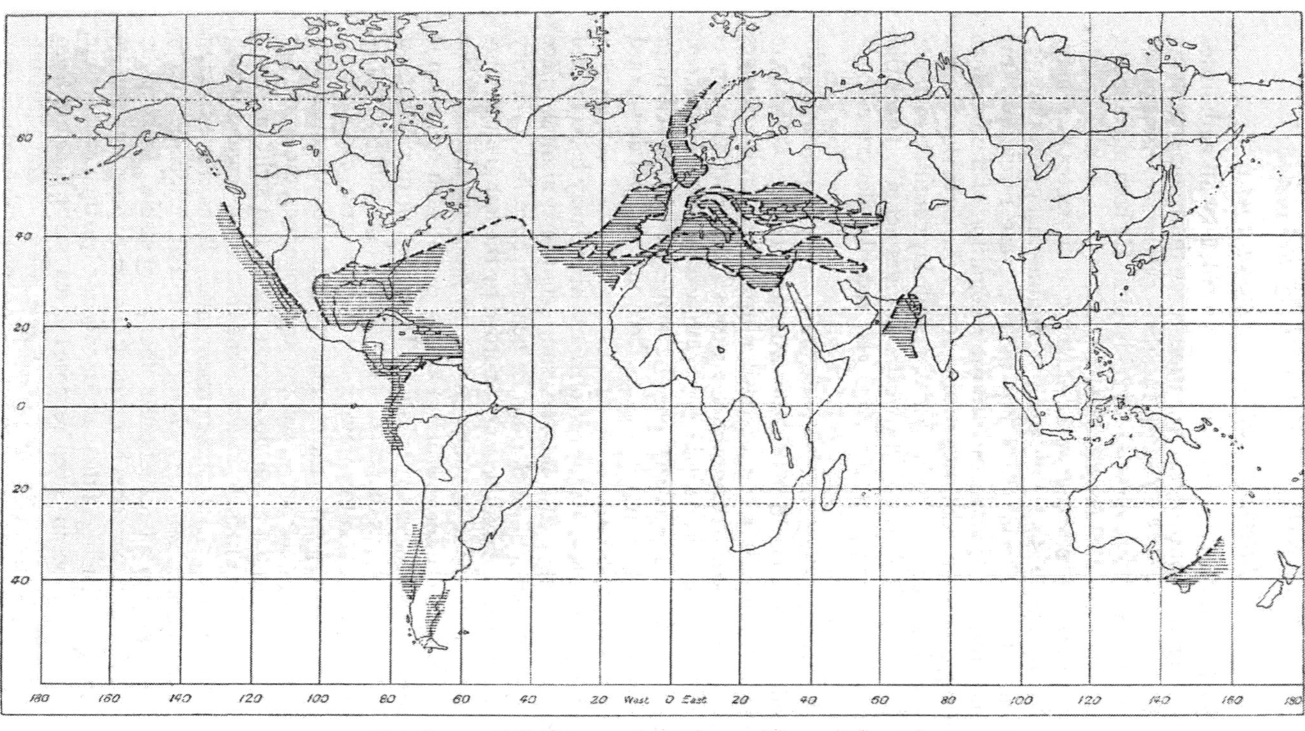


Esquisse partielle des mers helvétiennes (Upper Miocene)
After de Lapparent

# APPENDIX TO CHAPTER VIII

## *The Inter-Revolutionary Disturbances*

Inter-revolutionary disturbances are recorded by all authorities as affecting the continents. In some cases there appears to be orogenesis of a local character. In others there is an invasion of ocean waters on the lower continental levels. Or both phenomena may occur. To what may these be ascribed?

So far as these disturbances are orogenic in character they may, probably in many cases, be ascribed to volume-changes of the substratum. In the ages following upon a great revolution this explanation involves no more than the recognition of the fact that in the region just beneath the surface-crust of the earth slow movements, arising from final consolidation, probably continue for long periods. It is also probable that such changes actually affect the whole earth but only result in an orogenic record under crustal conditions which may be local.

Such surviving effects of a great revolution may be potential over a considerable period of time. Again we must recognize that, as the accumulation of heat in the substratum progresses, density-changes of opposite sign *must* begin, attended with all their own potentialities. Downward continental movements would first be expected. Then as the tidal creep began and the compensations shifted from one region to another in the magma, both upward and downward movements might occur.

Such possibilities as we have referred to above may be described as normal. But abnormal or cataclysmic events may also take place in the substratum. For the depths beneath have had a history which, notwithstanding the relief afforded by the last great revolution, may leave great volumes on the verge of instability. We cannot say that a lesser revolution arising out of such condition is impossible or even very improbable. It might run its course in a relatively few millions of years. Here again the results might be local and dependent on the prevalence in certain areas of such conditions as would respond to the comparatively feeble orogenic forces.

In connexion with inter-revolutionary orogenesis, the storage of energy in a deformed ocean floor must not be lost sight of. It is quite possible that this is the source of the vertical movements which affect the Japanese islands at the present time. It may have affected the Western Coast Ranges of North America at various periods in their history. For a buckled ocean floor is not stable. It keeps its position as an arch ring does by exerting continued pressure upon its abutments. A yielding of these abutments involves their cataclysmic overthrow. An ocean 'deep' exposed to magmatic pressure from beneath is an inverted arch and must give rise to lateral thrusting forces which, if the radius of curvature is great, might attain very great magnitude. Inter-revolutionary orogenesis involving lateral movements of a few miles might arise from these conditions.

## IX

## GEOLOGICAL TIME

Of much interest in tracing the surface history of the earth is the vexed question of Geological Time. This is often spoken of as the Age of the Earth—from which, probably much longer, period it is entirely distinct.

We shall not discuss the question of the age of the earth. For in fact, from what we have seen, the thermal manifestations on the surface of the globe give us no clue to it, even if we take for granted that the earth was once in a molten state from primitive heat. Whatever primitive heat the earth may once have possessed seems totally gone or at least reduced to insensible manifestations at its surface.

The case for the radioactive origin of terrestrial heat is a strong one. On the results of investigations on rocks from all parts of the earth, many of abyssal origin, we must logically conclude that radioactive elements are present throughout the rock-materials of the globe. There are no known exceptions. Further, all stony meteorites contain uranium. The one which has been investigated for thorium gave a positive result. In addition the study of haloes (presently to be referred to) affords evidence of the remote antiquity of radioactivity on the earth, and of the former existence of yet other radioactive elements.

We must conclude that if the earth ever was in a molten condition—which many regard as a reasonable hypothesis—it must have been at a period indefinitely remote.

On these grounds the view that the effects due to high temperature, so often revealed in surface phenomena throughout geological time, are of radioactive origin is clearly justified. Any other view involves gratuitous assumptions and the postulation of rock materials such as are unknown at the surface. Plainly, then, in considering the geological age of the earth we may claim freedom from any such limitations or inferences as might be based on its hypothetical cooling throughout geological time.

But modern advance has furnished other means of approaching the question of the duration of the geological history of the globe.

Thus we may proceed on the assumption that denudation has progressed during the past at an average rate not greatly different from that which now prevails. Then, by estimating the quantities of materials which have come into existence as products of denudation, and which exist at the earth's surface in forms capable of estimation, we can arrive at an approximation to the time required for their accumulation. We can apply this method, using the rate of sedimentary deposition and the estimated amounts of the sediments collected over geological time.

More definite in every respect is that method which uses the sodium in solution in the ocean as the modulus.[1] We may calculate the total mass of sodium in the ocean (an easily found and reliable quantity) and then apply to the rivers for information as to the rate at which this element is entering the ocean. The reason for selecting sodium from among the many elements in solution in the ocean is, that this element alone is not withdrawn from it by organisms, or in any appreciable amount otherwise abstracted.

Now all methods founded upon rates of denudation are in approximate agreement in ascribing to the most

[1] Joly, *Trans. R. D. S.* 1899.

ancient sediments an age which is about 100 million years. By making certain assumptions the sodium method may be stretched to 175 millions of years.[1]

Since the discovery of the radioactive elements an entirely different mode of estimating the Age has been evolved. It is based on the assumption that the final products of radioactive change in the case of the parent substances, uranium and thorium, are known. Both are, as it happens, metallic lead. The lead derived from uranium has a lower atomic weight than that derived from thorium. The former has an atomic weight of 206 nearly: the latter of closely 208. These atomic weights can be calculated on our knowledge of the atomic weights of the parent substances, and of the losses of mass which attend the several successive changes in the series of transformations resulting in the final stable element—lead.

Suppose we find a rock rich in uranium. We measure the quantity of uranium present and the quantity of lead which is found associated with the uranium. Next we infer from certain measurements the *rate* at which uranium is now changing into lead. Obviously, from the weight of uranium present and from the weight of the derived lead, we may calculate the Age. A similar procedure can be applied to thorium.

Of course, this method would not be reliable (*a*) if the lead was in either case unstable—i.e. ultimately changed radioactively into something else; or (*b*) were added to by such ordinary lead having the atomic weight of 207·2, as may be found in rocks; or (*c*) if the rate of change of uranium or thorium has not been constant over geological time, so that the resulting lead was produced at a faster or slower rate in the remote past.

It may be said that there is at the present time no direct

[1] See Appendix to this chapter.

evidence that, either in the case of uranium or thorium, the error involved in (*a*) exists. The atomic weight of uranium-derived lead has not as yet been found to agree exactly with its calculated value : 'whether the higher values' (found for uranium lead) 'are due to the presence of ordinary lead or to some unknown factor cannot be decided'.[1] In the case of thorium lead, a like small discrepancy has been found but of opposite sign. The suggestion has been made that thorium lead is not stable, but hitherto research has failed to find as consistently present in the ore the elements into which it might be supposed to be transformed ; or in any other way to support the suggestion. Errors from source (*b*) are guarded against by determining the atomic weight of the lead associated with the parent substance. Regarding (*c*) there is some evidence from another line of investigation that the rate of disintegration of uranium in early geological time may have been faster than now. It is difficult to explain the nature of this evidence in brief compass, but, as our considerations further on involve the question of the duration of geological time, and matters are also involved which are intimately associated with the earlier radioactive history of the earth, a general account will be included here.

In Chapter IV, referring to radioactivity, it will be seen that uranium and thorium give rise to a succession of substances which endure for a time ; these emitting certain radiations transform into other elements ; and so the *débâcle* proceeds. Of the radiations emitted, some (called alpha rays) consist of helium atoms carrying a charge of positive electricity, and possess a high initial velocity. This velocity differs according to the element from which they are derived. *And it has been found that*

[1] Fajans, *Radioactivity*, 1922, p. 49.

*the velocity is greater when the emitting element is short-lived and less when it is long-lived.*

In the course of the descent of uranium to lead, the slowest alpha ray is that which proceeds from the uranium atom itself when in process of transformation. In the case of thorium, the slowest is that which comes from thorium itself. This is in keeping with the very long average life of these elements.

Now if a small speck of a uranium-bearing or thorium-bearing substance is contained in certain coloured minerals of the rocks, these alpha rays will be found to have produced a darkening of the surrounding mineral. In the coloured micas (biotites) this effect is best seen. Of course, the fastest alpha rays get farther from the parent speck of radioactive substance than the slower rays. But even the fastest of the uranium series of rays gets no farther than 0·03 of a millimetre and the fastest of the thorium series no farther than 0·04 of a millimetre; and the slower rays penetrate each one a distance according to its initial velocity. As the darkening is accomplished mainly near the end of the trajectory of the ray the effect in the mica is to give rise to a succession of darkened spherical shells or surfaces concentric with the radioactive particle. Seen under the microscope in a thin cleavage flake of the brown mica, the appearance is that of concentric rings of darker colour. If the nucleus (visible as a central speck) contains uranium the outside radius will be 0·03 mm.: if thorium be in the nucleus the outside radius will be 0·04 mm. These circular markings have long been known to exist and are called pleochroic haloes, because of the optical changes visible with polarized light. Only in recent years have their complex structure and mode of origin been recognized.[1]

[1] Joly, *Phil. Mag.*, March 1907.

Without going into details, it may now be stated that when we endeavour to account for the structure of a halo as due to the combined effects of the several rays emitted from the central nucleus, in the case of the thorium halo we find we can do so with considerable accuracy. But in the uranium halo we find that the innermost ring of all as observed in the mineral is larger than it ought to be. Now this inner ring is mainly due to the alpha ray emitted by uranium itself, and when we recall that the range of the alpha ray has been found to increase with the shortness of the average life of the element emitting it, we seem led to the view that in remote times uranium transformed more rapidly than it does to-day. How might such a variation be explained? Only by the former presence of an isotope[1] of uranium which decayed faster than such uranium as we know of. This isotope must give rise to lead as its final product. It must break up considerably more rapidly than uranium such as has been investigated, so that it would, sensibly, die out in the course of geological time. We need not suppose that all

[1] Isotopes are atoms possessing different atomic weights but behaving chemically alike so that they cannot be separated by chemical means. They are specially dissimilar in radioactive properties. This is most conspicuously shown in their rates of decay. Thus uranium of to-day is known to contain two radioactive isotopes; one of which has a half-life period of four and a half thousand million years (that is, one half of it will have been transformed in this time), and the other has a half-life period of two million years.

The recent work of Aston has shown that isotopy is not peculiar to the uranium and thorium groups of elements, but that a considerable number (the majority) of those so far investigated are complex in the sense that they are constituted of atoms of distinctly different mass. Thus xenon, whose atomic weight is taken to be 130·2, is constituted of seven isotopes ranging in atomic weight (mass) from 128 to 136. Tin is composed of eight isotopes, zinc of four, and so on.

These important results emphasize the fact that the mass of the atom and its radioactive properties are referable to its nuclear structure, whereas its chemical properties are more directly dependent on its exterior features; the orbital circulation of electrons around the nucleus.

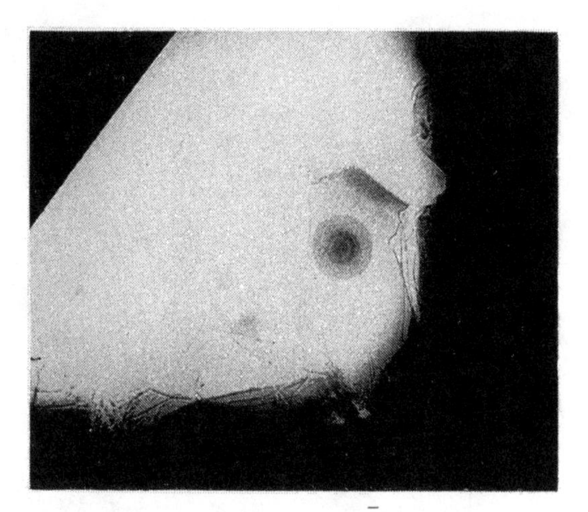

Thorium Halo in Björn Granite

Huronian (?) × 100.

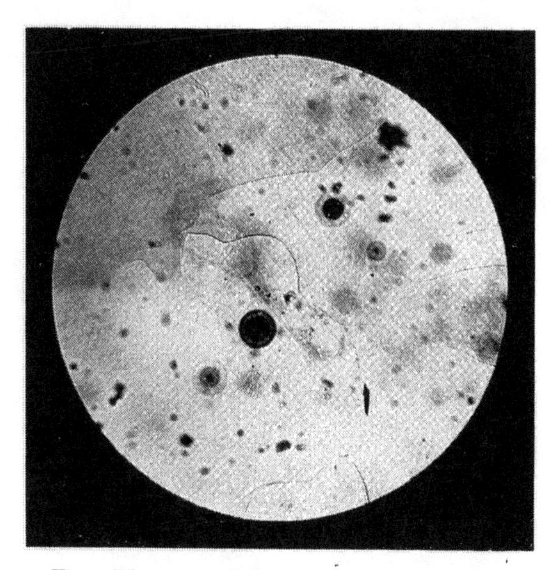

Four Uranium Haloes in various stages of development

From the Mica of the Leinster Granite-Devonian Age. Magnification 76 diameters.

Micro-photograph.

its products of change would be identical with those which we know of; although to account for the fairly good fit of the outer features of the halo these products must possess similar or approximately similar ranges.[1]

Quite recently all this has been confirmed so far as halo-measurements can confirm it. Measurements made on some beautifully defined uranium haloes in fluorspar, by Gudden, show that the alpha ray of uranium must formerly have possessed a range which (in air) would have been 2·68 cm. instead of 2·53 cm. as deduced from recent measurements, and, further, the succeeding ring did not possess a radial dimension such as the next derived substance ($U_2$) should exhibit. The range (reduced to air) was 2·76 instead of 2·91 (that of $U_2$).[2] In fact the uranium responsible for the primary ring decayed about fifty times faster than recent measurements indicate, and the succeeding derivative decayed fifty times as slowly, if the law connecting range with longevity be true.

The matter cannot be fully discussed here. Unless some other explanation can be found, there is no question that a serious uncertainty is introduced into the value of uranium-lead measurements by the now fully confirmed fact that the primary range as recorded in ancient haloes is longer than is now observed and would be expected from its present rate of decay.[3] This misfit was first found in haloes of Devonian age. There is evidence that in haloes formed in mica of late Tertiary age (the Mourne granite) the misfit of the primary halo-ring seems less, and in mica of Archaean age (Ytterby mica) it seems to be greater than what was found in the case of the Devonian

[1] Joly, 'The Genesis of Pleochroic Haloes,' *Phil. Trans. R. S.* 217, pp. 77–8, 1916; and 'The Age of the Earth', *Nature*, Jan. 15, 1922.

[2] Gudden, *Zeitschrift für Physik*, Aug. 1924, p. 110.

[3] The reader is referred to Rutherford's *Radioactive Substances and their Transformations*, pp. 447 and 607.

haloes. This would, of course, support the explanation suggested above.[1] If the offered explanation is correct it would follow that the chronological reliability of uranium-lead ratios diminishes with the antiquity of the rock.

The following passage, cited by Ellsworth[2] from a communication received by him in 1922 from Sir Ernest Rutherford, speaks for itself. Its importance as bearing on the matter under discussion is obvious.

> 'Personally I am not at all sure whether we can rely entirely on our estimates of the half periods of uranium and thorium by counting alpha particles; for we do not know the composition of uranium and thorium from the point of view of isotopes. It may be that the activity of uranium is mainly due to one isotope which is a fraction of the whole. In addition it is possible that isotopes of shorter life were originally present and have died away. As you may know, there has always been some uncertainty about lead and thorium as the ages from the lead ratio do not agree with other data.'

What now are the readings derived from uranium and thorium lead for the age of the rocks? According to ages determined from the uranium-lead ratio, the Devonian period would be about 325 millions of years ago. For the Pre-Cambrian, ages extending from 560 to 1,340 millions of years have been read.

Thorium-lead ratios give much lower readings. Lead derived from a Norwegian thorite revealed for the age of lower palaeozoic rocks 150 millions of years. The atomic weight of the lead was in this case investigated and found to be that of thorium lead. A selected specimen of thorite from Ceylon afforded an age of 130 millions of years, whereas a uranium-lead ratio gave for these rocks 512 millions of years; that is about four times as great an age as thorium lead indicates. If thorium-lead ratios are trustworthy we should divide the

[1] Joly, 'Pleochroic Haloes of Various Geological Ages,' *Proc. R. S.*, vol. cii, 1923, p. 694.

[2] *Am. Journ. of Science*, Feb. 1925.

uranium-lead ages, as determined for very old rocks, by four.

We seem to have reached a conclusion entirely in favour of the lower estimates of geological time. According to estimates based on oceanic sodium, the time since the products of solvent denudation began to collect in the ocean may be as much as 175 million years ago. The thorium-lead ratio would probably make it 250 millions of years. The approximation—rough though it be—is striking when it is considered that there is not one single factor common to both methods of calculation.

In the present connexion we have to face the question whether any contribution to the subject of geological time may be forthcoming from such time-estimates as may be attempted respecting the cyclical events of earth-history. Now what we have to go on is—at best—rather indeterminate. The period required for the latent heat of basalt (as we estimate it) to be accumulated and fusion to be brought about is some thirty-three millions of years on the higher value of the radioactivity of plateau basalts, and is, possibly, fifty-six millions of years if the lower values prevail in the substratum.

The rate of heat loss is even less definite. It turns for a large part on the limiting thickness of the ocean floor and on the depth of the substratum which takes part in a particular revolution. The latent heat collected beneath oceans and continents to a depth of 32 km. (20 miles) beneath the latter would escape through an ocean floor 6 km. (3·8 miles) thick in 5 million years. It is possible that the floor is more reduced in thickness. On the other hand, it seems probable that much greater depths of the substratum are in general involved. Finally, we have to add to these considerations the certainty that a large part of the heat will be discharged by the direct escape into the ocean of the fluid magma through dikes and rifts. It would seem to be sufficient to assign 5 million years to the period of thermal release.

We are therefore confronted with estimates which may involve forty millions of years as probably a low estimate and sixty as probably a high one. Of course, much is involved in these figures which is very imperfectly known; more especially the physical data under conditions of high pressure.

None of the authorities referred to in the last chapter estimate more than five *complete* cycles as entering into geological time. Sonder's estimate would involve four, Barrell and Schuchert would say five.

Finally, then, our estimates leave us with from 200 to 300 millions of years as the duration required by the great cycles of earth-history. But according to some authorities the limits would lie between 160 and 240 millions of years—that is, if four cycles fill the whole period of the earth's surface history since the Laurentian Revolution.

These figures are most nearly in agreement with the lesser estimates of geological time.

## APPENDIX TO CHAPTER IX

### *Finding the Age by Solvent Denudation*

The method of finding the age of the oceans by the sodium content of the oceans has been the subject of considerable discussion. A brief reference to the principal points that have been raised will be given because it will be found that its testimony, after all is said, retains its value.

1. It is suggested that existing conditions tend to minimize the Age, because we live at a period when there are no transgressional seas to cover the lower continental levels; such as prevailed over long ages in the past and must greatly have diminished the activity of surface denudation.

In answer it is contended that the mere extent of the land surface does not, within limits, affect the rate of denudation. The existing rain supply is quite insufficient to denude the whole land surface, about 30 per cent. of which does not drain to the ocean. If the transgressions covered 30 per cent. of the continents, the rainless area may be supposed to diminish and the denuded area remains the same, seeing that no change in the rain supply, save one of increase, is to be expected. There might, even, be increased subaereal denudation under these conditions. Existing rainless areas show every sign of having been exposed to active denudation in past times.

Further, as some guide to the average continental area in the past, Schuchert's estimate that the average area of North America throughout geological time has been about eight-tenths of its existing area may be adduced.[1]

Again, it has been found by comparing the solvent denudation as estimated for the different continents, that there is no connexion between continental elevation and subaereal denudation. Europe, the lowest of the continents, delivers the most matter in solution into the ocean; and North America and Asia, although their average heights are very different, deliver nearly the same amounts of dissolved matter into the seas.[2]

2. It has been said that the chloride of sodium in the ocean may be carried inland by winds, and in this manner circulate from sea to land and back again, and so falsify the true denudative supply of the rivers. The answer is that an allowance can be made for the

[1] *Bull. Geol, Soc. Am.*, vol. xx, 1910.

[2] Sollas, *The Age of the Earth*, Fisher Unwin, 1905; Joly, *The Birth-Time of the World*, Fisher Unwin, p. 1 et seq.

*limiting* effect of this error by assuming that *all* the chlorine in the rivers is derived from the ocean and carries with it its due proportion of sodium. This would raise the geological age to $141 \times 10^6$ years.[1]

3. Salts derived from the ocean might be retained in the sedimentary rocks and unduly increase the river supply. The answer is very simple. We must suppose these assumed supplies to enter the rivers according as the rocks are denuded. The rate of denudation therefore controls this supply. Calculations based on this rate and on estimates of chlorine in such rocks show that the Age would be affected at most by 0·9 per cent.[2]

4. A possible primitive acid denudation by condensing gases might have introduced into the ocean an initial amount of sodium. It can be shown that this effect is probably small. As it would reduce the numerator in the fraction

$$\frac{\text{quantity of sodium in the ocean}}{\text{quantity supplied by rivers}} = \text{Geological Time},$$

the effect is, of course, a negative one; unduly diminishing the derived age.

Sollas, as the result of a careful examination of all the data, arrived at extreme limits of 80 and 175 millions of years.[3] Some geologists accept the great ages derived from the uranium-lead ratio; but no explanation has been forthcoming as to how the solvent denudation now progressing on the earth can be eight or nine times greater than the average over the past.

The biological evidence as it bears on this question has been discussed by Sollas and should be read.[4]

[1] *Radioactivity and Geology*, Constable, p. 245.

[2] *Geol. Mag.*, vol. vii, May 1900, p. 220, and August 1901, p. 344; also *Report Brit. Assoc.*, 1900.

[3] Presidential Address to the Geol. Soc., vol. lxv, May 1909, p. cxii.

[4] *The Age of the Earth*, Unwin, 1905.

# X

## THE DOMINANCE OF RADIOACTIVITY

WE shall now fill in some details of our history, crediting the reader with a recollection of those broad features of the surface structure of the earth upon which we have based it, and in which, as we have contended all along, that history inevitably originates.

What do we know about the veritable beginning? We may answer nothing definitely. It is believed by some that the earth originated in a very improbable and fortuitous event—a collision (or close approximation to one) between some wandering body and our primeval sun. And after that a very long period of physical changes at last gave us the world somewhat as we know it.[1]

Isostasy must have been an early feature of earth-history. We imagine the light continental rock-magma collecting scum-like on the surface of a basaltic lava covering the entire earth. Flotation was inevitable ; as inevitable as the flotation of ice forming on the surface of the sea. All along the heat may have been mainly radio-active in origin—or at least for aeons before geological history began.

Antecedent to the genesis of the moon—if we accept Darwin's beautiful and convincing theory of its origin—Geological History as we know it was probably still for the far future.[2] For, as we have seen, we owe to tide-raising

[1] See Jeans, *The Nebular Hypothesis and Modern Cosmogony*, the Halley Lecture, 1922.

[2] The theory referred to assigns to the moon a terrestrial origin at a period when the primitive earth was rotating axially at a velocity which may have been once in from 3 to 4 hours. At this speed of rotation the semi-diurnal tide raised by the sun would possess a period of from $1\frac{1}{2}$ to

forces, coming into operation when the substratum is liquid, the persistence of the continents. When the moon formed part of the earth, tide-raising forces were restricted to those due to the Sun: now about 3/7ths as great as those due to the moon.

The persistence of radioactivity over the entire span of geological time finds clear proof in the study of certain haloes as referred to in the last chapter. These are found abundantly in the ancient black micas of Ytterby which are of Archaean Age. The appearance of these haloes is indicative of their very great antiquity. They differ greatly from haloes observed in mica even of Lower Palaeozoic Age. They have been changed in the same sort of way that an over-exposed photograph is changed. They are, in fact, 'over-exposed'—or 'solarized'. Now there appears to be much in common between the manner in which the alpha radiation affects the mica and the manner in which light affects the photographic plate. It is well known that a very long exposure to light will reverse the latent image, so that what should develop as high lighting behaves like the unexposed plate. In the genesis of the halo the same effects occur. The most ancient haloes are bleached just where they should be darkest. Sometimes they are so

2 hours. Now this period would correspond with the free period of oscillation of the earth, assuming it to have been a homogeneous liquid throughout.

We have then an applied force keeping time with the natural or free period of oscillation of a material system, a condition which must result in ever-increasing amplitude of movement. The tidal oscillation in this case might have become so violent as to result in the disruption of the planet, so that great fragments became detached which ultimately coalesced to form the moon. The genesis of the moon was on this theory a phenomenon of resonance such as may be illustrated by a familiar example—the maintenance of the motion of a heavy clock pendulum by a small force applied in time with its period of vibration (Darwin, *The Tides*, p. 254 et seq.).

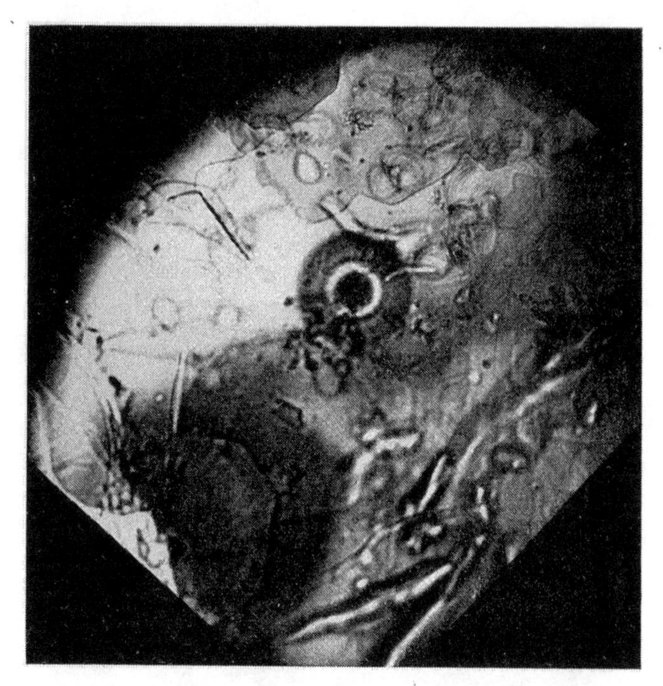

Halo of unknown origin. Ytterby Mica (Archaean) × 256
Micro-photograph

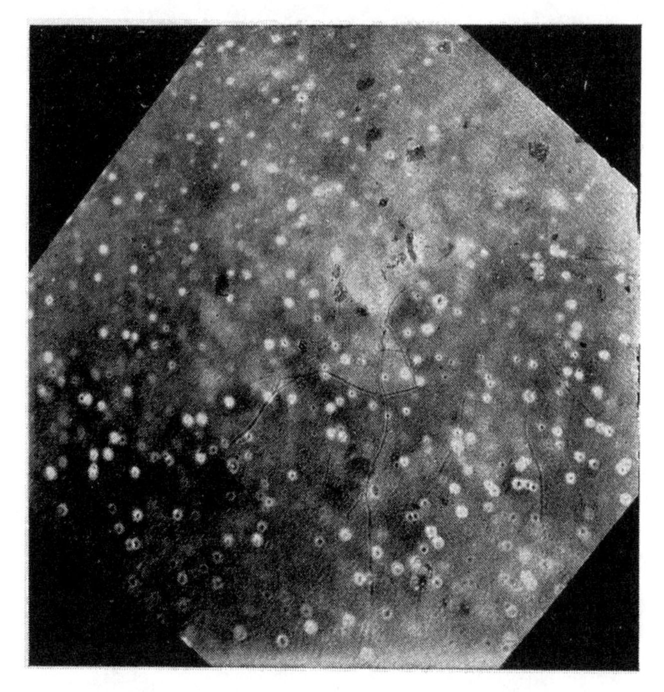

Haloes of unknown origin in the Mica of Ytterby (Archaean) × 106.4
Micro-photograph

much over-exposed as to be bleached altogether. These appearances, whatever be their origin, are conspicuous in the most ancient haloes. We may recognize an Archaean halo by these characteristics. They prove the activity of the parent radioactive element—whether it be uranium or thorium—even in those very remote times. Sometimes in these micas we meet haloes which appear to have been formed by radioactive elements no longer existing. These haloes seem to be unknown in younger rocks.[1]

A substratum so charged with energy as that of the earth can never be at rest. To-day, at the close of a great mountain-building period, deformative movements are still progressing. Hence the faulting and dislocation which in many cases appear to be the origin of seismic phenomena. And we find that the great earthquake regions of the globe are just those tectonic areas in which mountain building more especially prevailed. Are these crust-movements purely residual and reminiscent of the recent orogenesis, or are they the result of the accumulating radioactive heat within the substratum? We cannot tell. For either the one or the other might be credited with the origination of faults and dislocations. But the former view is supported by much *prima facie* evidence. On the eastern and western shores of the Pacific seismic phenomena proceed almost continuously, and we know that these regions—both continent and ocean floor—were lately affected by immense crustal stresses. The Atlantic shores have not been affected by great recent stresses, and they exhibit no marked seismic phenomena.

When we bear in mind the greatness of the forces which have affected the surface crust of the earth within

[1] Joly, 'The Radioactivity of the Rocks,' *Trans. Chem. Soc.*, 1924.

geologically recent times—the evidence for which is proclaimed by all the great mountain ranges of the earth—we find no difficulty in tracing seismic phenomena—relatively trivial in magnitude—to the radioactive history of the substratum. Deep within the substratum itself movements resulting in seismic radiations also occur. Our views as to their origin and nature must, at the present time, be speculative.

When we study geography we do not generally give much thought as to how oceans and continents came to be as they are. Nevertheless, if we consent to trace back the physical history we shall see that the proportion of land to water on the globe is as it is, mainly because of the deep, fundamental structure of the earth's surface. We shall find, in fact, that not only has the system of the earth, as outlined in preceding chapters, been a dominant factor in shaping the events of geological history, but it has left its record upon the larger geographical features of the globe.

The relative areas of land and water are clearly determined by factors such as we have already recognized as operating in other directions. We saw that owing to its proper radioactivity the continents at their base must approximate to the temperature of melting of the underlying substratum. (We here speak of the *average* basal temperature of the continents.) Now this basal temperature increases as the *square* of the depth of the continental layer. Thus any considerable change in the depth involves a large rise in the basal temperature. But if this temperature were to be generally much greater than it is, the stability of the compensations must be jeopardized; for heat must then escape downwards into the substratum. This becomes raised to a temperature above its melting-point (the hotter lava necessarily gravitating upwards), and towards the close of the long period of heat-storage the continental rocks themselves must soften and melt and flow away laterally—probably eastwards—when tidal movements come into operation. As we have seen in foregoing pages, the compensation of the Tibetan Plateau must approximate to those unstable conditions.[1]

Upon the upper continental surface effects of denudation operate in the same direction, transporting materials from the heights into the coastal waters and so spreading laterally the margins of the continents.

In these effects we find factors tending to reduce the thickness of the continental layer.

Operating in the opposite direction we have the lateral crushing of the continents which arises out of the periodic increase and diminution of the earth's radius. Plainly if these lateral forces continued to act indefinitely the continents must ultimately come to occupy an ever-lessening area upon the surface of the basaltic substratum. But the opposing factors are at work. The continent which has grown too thick, denuded both above and beneath, sheds off some of its mass ; increasing in area, diminishing in depth.

These opposing physical conditions, which have been in operation all along, control not only the area of the continents, but, necessarily, also that of the oceans. And, we may add, also the average depth of the oceans as

[1] It is a fact that many of the volcanic islands of the Western Pacific exhibit problematic continental rocks along with the basaltic and other eruptives of which essentially these islands are composed. The distribution of these continental materials is described by Daly (*Am. Journ. Sc.*, February 1916, p. 161). It is worthy of consideration whether they may not have reached their destination by ejection from the subcontinental materials of Eastern Asia, and the movement of the ocean floor relatively to the underlying magma ; representing, in fact, the results of igneous denudation progressing in the underworld.

determined by the volume of water upon the surface of the globe.

The variation of climate and origin of 'Ice Ages' over geological time raise many interesting questions. There is good evidence that

'cold climates fall in with times when continents were more or less extensively and highly emergent. There were no cold climates when the continents were flooded by the oceans.'

On the other hand, mild climatic conditions, e. g. such as existed in Upper Carboniferous times and Upper Cretaceous and which nourished the rich floras of the Coal Measures,

'accompanied the time of greatest continental flooding and preceded the appearance of cooled climates'.[1]

Prof. Ramsay of Helsingfors connects changes of climate in geological time with the general rhythm of geological cycles. Severe climatic conditions attend the high relief prevailing during and after the great orogenic periods, warmer climates following the erosion of continents and mountains. Glaciation arises as a result of maximum relief. Interglacial periods do not imply corresponding fluctuations of relief but depend on variations of sea-level brought about by crustal movements.[2]

According to Haug, the study of the oscillations of the Scandinavian shield has shown the relations which exist between epeirogenic movements[3] and glacial phenomena. The uplifts lead to glacial caps which invade the whole area of uplift. The depressions give rise to melting of the glaciers and their gradual retreat. Generalizing on this deduction, a general cause can be

[1] Schuchert, *Rep. Smithsonian Inst.*, 1914.
[2] *Geol. Mag.*, 61, 1924.
[3] Movements of uplift tending towards continental growth.

Two views of the Zinal Rothhorn (13,855 ft.)

The upper view is from the Mettelhorn (11,188 ft.) showing the east side; the lower, from the Corne de Sorebois (above Zinal), shows the north-western face of the mountain. Both display strikingly the effects of glacial denudation.

assigned to the phenomena of the extension and retreat of glaciers. The synchronous phases of glaciation over the globe must be attributed to the synchronism of positive [upward] epeirogenic movements.[1]

It would appear from this that the changes in vertical elevation of the land are responsible for climatic variations. This is in accordance with the view that the atmosphere and its water vapour exert a 'greenhouse effect' upon our climates. Luminous solar radiation is freely transmitted. But this when absorbed at the earth's surface and converted to heat-radiations is less freely transmitted outwards. It is absorbed in and conserved by the atmosphere and by its moisture. The effects arising from this are similar to those which warm the greenhouse, the sun's radiations being unable to return by the way they entered.

The climatic conditions, therefore, largely depend on the depth and moisture of the atmosphere covering the continents. Hence when the substratum is fluid and the continents depressed relatively to the oceans, all the conditions for warm and moist climates exist. When the continents are elevated the conditions are cold and dry, similar to those obtaining in the high plateau regions of the earth. And in this way Ice Ages appear, simultaneously affecting the whole world in times of great emergence.

In our school-days we were told that the surface features of the globe are very inconspicuous when contrasted with its great dimensions. The surface rugosities were compared to those of an orange. But, in truth, they are still smaller. For if we consider a model of the globe, one metre in radius, the average terrestrial inequality of

[1] Haug, *Traité de Géologie*, i, p. 501.

level—i. e. the height from the average depth of the ocean floor to the average height of the continents (approximately 3 miles) on our model globe would hardly stand higher than the thickness of a visiting-card ($\frac{1}{4}$ millimetre).

We see now how this comes about. The bulk of the continental rocks is so much, and the bulk of the waters is so much. And periodically acting forces tend, some to enlarge, others to restrict, the area occupied by the continents on the basaltic surface of the globe. The final areas of land and water as we see them are the results of the contending influences.

But in the course of these periodic adjustments the continents get pressed into wrinkles. The heights of these wrinkles are controlled by isostasy, and by the fact that compensations which are too deep are unstable and melt away beneath. Hence the scale upon which rugosities can exist upon the earth's surface is defined largely by the stability of the compensations. However, the mountains may spread their load over areas wider than those they occupy at the surface, and can thus to some extent evade the restrictions imposed by the instability of the compensations. If the Himalayan peaks had to be individually compensated, it is certain they would not be stable: their compensations would melt away and the mountains sink down. It is possible that in certain cases such a failure of the compensations may account for under-compensated areas as detected among these great mountains.

But the mountain heights are also controlled by detrital denudation which probably increases with the height of the mountain. This wasting away of the lofty mountain must give rise to over-compensation—i. e. the compensation which was right for the higher mountains is too great for the lesser—and the mountains tend to rise.

This condition also appears to have been detected in the Himalayas.

Thus we can easily explain why our mountains are so relatively small, and we can refer the height of the land and depth of the ocean to the operations of general laws controlling the equilibrium of the materials floating on the surface of the great and universal substratum.

Fig. 10. Sector of the earth to a scale of one metre to earth's radius

We can advance yet a step farther. We can perceive a reason for the broad geographical features of the globe, that is, for the general disposition of continental and oceanic areas upon the earth's surface.

What are the salient features of terrestrial geography? The answer is: *Continents sundered by seas which extend from pole to pole.* Now there are forces, feeble it is true, but always acting, tending to urge the floating land masses towards the equator. These forces arise out of the rotation of the earth on its axis: a slight excess of centrifugal force acting on the continents because of their higher 'freeboard' and correspondingly higher centre of gravity. It may be that these forces are negligibly small, although, as stated in Chapter I, they have been appealed to as in part causative of continental movements. However, it is not hard to show that any aggregation of the continents along the equator so as to form a band or girdle around the earth would not be stable.

This appears when we recall that the tidal shift of the outer parts of the crust, whereby the ocean floor is brought over that part of the substratum formerly occupied by the continents, offers the only avenue whereby the subcrustal magma can part with the latent heat which for long ages has accumulated therein.

Now this means of release would absolutely fail if the continents extended in continuous belts, or were merged into one continuous belt, encircling the globe. There could be, in that case, no escape of heat into the ocean. What is more, the tensile effect directed east and west and arising during the period of magmatic expansion, would fall entirely upon such a continental belt, there being no interposed sea-floor to absorb the tensile forces. Both conditions must contribute to the rupture and break up of continents so distributed. Any large polar continent would, for similar reasons, be unstable.

Lastly, we shall briefly consider the bearing of the cyclical history of the globe upon the progress of organic life.

And first we may ask, 'What is the living organism?' Now although the living thing is in detail inextricably complex, in its purely physical aspect and from a dynamic point of view it is capable of being defined. The organism may be regarded as a mechanism which—for purposes of growth and reproduction—absorbs and utilizes energy *acceleratively* from whatever available source the energy can be derived. This attitude is seen in the advancing demands of organic life in all its forms upon the energy of the environment and—taken in conjunction with the fact that the energy supplies are limited—constitutes the dynamic basis of evolution. If we reflect on this statement we shall find that it differentiates life from all other material configurations whatsoever upon the surface of the earth.[1]

Now in the foregoing pages we have seen that all over the earth's surface great physical changes take place after long periods of slowly developing continental seas. How do these changes affect organic life?

We all know how wonderfully the organism responds to environmental changes. This is brought home to us as we travel over the world in the fact that we find plants and animals everywhere adapted to the climatic, physical, and organic circumstances of their environment. For, indeed, specialization to meet the environment arises out of the dynamical conditions controlling organic life, to which we have referred. For through specialization, principally, fresh energy-supplies are made available to the organism. In the coming and going of marine conditions upon the land, new restraints are imposed which benefit some and destroy others. Air-breathing life in the ocean was probably initiated in this way. And, again, marine forms abandoned by slowly shrinking seas developed air-breathing forms upon the land.

In a word, we conclude that, had our world been monotonously free from change, save diurnal and seasonal ones, much of the diversity of animal and vegetable life would never have been.

But while increase in variety must have arisen out of terrestrial surface changes, at the same time other consequences followed which might, in a sense, seem to have led to opposite and opposing effects. For the more widespread surface changes undoubtedly promoted uniformity of organic development over the earth. More especially does this apply to comparatively lowly forms of life, for instance, to the bivalve shell-fish which as lamelli-

[1] Joly, 'The Abundance of Life,' *Proc. Roy. Dublin Soc.*, vii. 1890.

branchs or brachiopods date back to very remote times. These, by the oceanic transgressions, became uniformly and widely distributed over the earth. Genera from the Gulf of Mexico mingled with those of the Arctic Ocean ; forms from the Atlantic found in the Tethys a highway to the Pacific. They swarmed into the epeiric seas and left their fossil remains in the muds which later built the mountains. They retreated again into more primitive seas—in the vastness of which they found comparative freedom from the struggle for existence—when the transgressional waters disappeared. Thus their development over the earth became equalized.

And this equalization was not only in space but in time. For the synchronous revolution, as it were, beats time over the whole earth to the progress of life upon the globe. What happens in America also simultaneously happens in Eurasia. And hence the spatial community and synchronous evolution of those organisms which have most contributed to our power of timing the events of earth-history over the globe.

Life so delicate, so frail in the individual, so strong, so immortal in the race, must have been very different from what it now is, had that trace of the radioactive elements been wanting in the rocks.

We can best realize what that trace of radioactivity means to the life upon the earth by looking forward to a day when it will at length be worn out. Mountains, unrejuvenated, must then sink down into the plains. Continents worn away age after age by sea and sky must be washed irrecoverably into the ocean. Air-breathing life upon the land and land vegetation must finally perish. For the earth itself will have ceased to breathe. And the mind of Man, which alone comprehends it all, will have become part of the forgotten past.

# APPENDICES

## I

## EFFECTS OF TIDAL FORCES UPON OROGENESIS

As stated in the foregoing chapters, it is recognized in tidal theory that the tidal wave which is maintained on opposite faces of the globe must act as a brake applied to the surface of the earth. This brake, operating to retard the west-to-east motion of the earth's surface, may be expected to give rise to a motion of the whole outer crust of the earth with reference to the interior—that is, if viscous movement of the outer over the inner parts be possible. In this way the length of the day may not represent the real rate of rotation of the earth, which spins within the outer crust with slightly greater angular velocity. Darwin [1] looked to such effects, acting in earlier periods of the world's history upon a viscous planet, as possibly wrinkling the surface in such a manner as to mould the observed continental contours. Eddington [2] considers it possible that purely oceanic tidal movements, brought to rest in land-locked areas, might have given rise to crust-wrinkling.

The tide-generating force is intrinsically a feeble force at the present time. The horizontal tidal force is only one eleven-millionth part of the weight. In past times, according to the tidal theory of the moon's history, the force must have been far greater, the horizontal tidal force increasing inversely as the 6th power of the moon's orbital radius.

According to the theory we have been discussing, at intervals in the world's past history there came into existence a sea of molten rock having a depth probably not less than twenty times that of the ocean and possessing three times the density of water. In its upper parts this lava-ocean was, probably, highly mobile—almost as mobile as water. The question arises, how far may tidal movements prevailing in such an ocean have contributed towards crustal disturbances in the remote past, and how much would they be effective even in Pliocene times or in the future? Data as to the downward viscosity of the magma ocean are not available, and doubtless they must be important factors in determining the answer to these questions.

It seems, however, certain that the retardative effect referred to must be influential chiefly in the upper regions of the magma ocean, for its lower layers, in contact with more rigid materials beneath,

[1] *The Tides*, p. 276. [2] *Nature*, January 6, 1923.

and, owing to pressure, probably possessed of greater viscosity than the upper layers, must partake more nearly of the full angular velocity of the earth's rotation. It is easy to see that these conditions must give rise to a west-to-east drive or pressure of the heavy magma against the immersed parts of the continents. It must be remembered that this state arises at a time when the ocean floor has been reduced in thickness, so that the submerged coasts of the continents are left exposed to the effects of differential movements in the magma. For, in fact, the horizontal tidal forces exert an east-to-west pull upon the continents, and this is resisted by the earth's rotational energy transmitted through the medium of the molten magma.

These conditions involve differential effects upon the floating land-masses. For the more deeply compensated continental features experience the west-to-east magmatic drive more than shallower land-masses. And, again, turning or rotational forces would arise, due to those greater downward-extending compensations which might be located excentrically respecting the resultant or centre of magmatic forces, or where continental masses extended partly into equatorial areas—the linear velocity of the magma being greatest at the equator. Such rotational effects may have affected mountain development. Comparatively small movements would suffice : thus, if the African continent and Peninsular India were, even to a small extent, affected in this way, by the more active equatorial currents upon their southern extension, Eurasian folding must have ensued. We have seen that at the height of revolution the ocean floor is probably much disturbed as well as reduced in thickness, and its resistance thereby diminished.

The effects of these conditions on mountain-folding would probably be principally experienced where the geosynclines had forced the continental materials deep into the magma. Here the east-going magmatic pressure would be largely concentrated and the temperature conditions also most severe ; lateral folding, and depression into still greater depths, being a probable result. Even under pressures of no very great intensity per unit area, provided they were long continued, these results might be expected. The final buoyancy effect would ultimately force upwards, isostatically, the bent and folded masses—along with batholithic invasions of the basaltic magma. Thus the ranges raised above the general continental levels would be formed.

It may well be that whereas both the west and east coasts of the Pacific owe to an expanded ocean floor the major wrinkling of the continental surface, the predominant orogenic phenomena displayed on the eastern coasts is due to a force which cannot have acted upon the western coasts—the magmatic pressure arising in horizontal tidal and precessional forces.

Schweydar[1] has investigated a precessional force which tends to urge the continents westward, and which has, of course, like the tides, an astronomical origin. From his statement it appears that this force must be considerably greater than the horizontal tidal force, and, obviously, its effects must greatly augment in times of revolution. It is strongest at the equator and is null at 36° North and South. It must conspire with the tidal force to retard the west-to-east movement of the earth's crust.[2]

It may be significant as regards this probable intervention of tidal and precessional effects that many of the great lava-floods of geological history have welled up through fissures on the western sides of continental land or west of great mountain ranges, or, again, have appeared where rifts ranging north and south have been attended by deep crustal faulting and sinking. The Deccan basalt thickens towards the west, and western coastal fissures have been traced. The Hebridean basalt was poured out westward of a continent. The Disco basaltic flows appeared to west of the Greenland ranges. The western inter-montane regions of North America are largely flooded by basaltic outflows. The great African rift appears to have determined intense volcanism and basaltic outflows. The Christiania rift seems to be a similar case. The existence in revolutionary times of easterly basaltic pressure beneath would account for this remarkable localization of igneous activity.[3]

Tide-generating forces acting on the continental crust have, according to the views of Kosmatt, been responsible for the meridional tendency of the greater rift valleys of the earth. It is probable that the tidal forces did actually determine this remarkable meridional tendency. The rifts were formed at a time when the substratum was fluid and the crust in tension. The crust tension acts, of course, in all azimuths, but a tidal force acting latitudinally may well have determined the direction of the fracture. We find support for this view in the fact referred to by Gregory that the African Rift is best developed in equatorial regions.[4]

[1] *Zeitschr. d. Ges. f. Erdk. zu Berlin*, 1921, s. 120–5.

[2] See also Wegener, *Die Entstehung der Kontinente und Ozeane*, Berlin, 1922.

[3] *Phil. Mag.*, June 1923.

[4] Gregory, *The Rift Valleys and Geology of East Africa*, p. 374, Seely, Service & Co.

## II

## THE QUESTION OF CONTINENTAL DRIFT

The possibility of the continents having shifted their relative positions during geological time has been urged by Taylor in America and by Wegener in Europe.

That such movements can have taken place while the substratum is solid must be regarded as improbable in the last degree. Nor have any forces adequate to overcome the resistance of the ocean floor, even during times of magmatic fluidity, been hitherto adduced. If, however, we may suppose that the ocean floor under certain conditions had become greatly attenuated or rifted in favourable directions, then it seems quite possible that the stresses arising out of tidal forces, as referred to above, might have sufficed to create differential motions among the continents. For, as pointed out in Chapter III, the draught of the continents is by no means equal. And we may picture the west-to-east drive of the magma, acting during long ages on the great compensations of Eastern Eurasia, sufficing to shift the whole great continent relatively to America. And, again, the deep compensations of the Andes may have been effective in conferring upon the southerly continent a position to the east of the northerly continent of America. Mountainous New Zealand may, similarly, have been carried eastward from Australia. Tensional rifts and faults extending north and south, generated during periods of increasing liquefaction, might initiate and define the new geographical delimitations. Rifting and faulting of the coastal ranges of Eastern Asia have undoubtedly been concerned in the genesis of the Japanese islands.

# A SUMMING UP: THE SURFACE CHANGES OF THE EARTH

A lecture delivered to the Geological Society of London on May 2, 1923

It is well known that the great catastrophic events of Geological History are cyclical in their succession. At intervals, measured by many millions of years, the continents exhibit a downward movement relative to the oceans. The seas transgress, covering, often, the major part of what before was dry land. These transgressions are preceded by periods of fluctuating invasion over the more low-lying areas. Other signs of crustal disturbance foretell the final great submergence. In the transgressional seas so formed sediments may accumulate to depths of thousands of metres.

Then, finally, after long ages, a time comes when all is reversed. The seas retreat slowly, mountains arise where sedimentation had been most profound. The former downward movements not only give place generally to upward movements, but, locally, those parts of the crust—the geosynclines—which had been most depressed by the burthen of deposits, become uplifted into the heavens, high above the general continental level. Such sediments come up folded, overthrust, and often profoundly metamorphosed.

Preceding the climax of this extraordinary series of events, floods of basalt arise and flow out through cracks in the earth's surface.

In the decipherment of these impressive features of terrestrial tectonics we owe much to the observational work and clear vision of American geologists. They have named the great mountain-building paroxysms 'Revolutions'. Both in the North American continent and in Europe several such revolutions have been recognized as having occurred since Cambrian times. One is inferred as having occurred in Silurian and Devonian rocks (the Caledonian Revolution); a second occurred in the Permian (the Appalachian Revolution); a third, on a great scale, in the Cretaceous and early Tertiary, responsible for the major development of the Cordilleras of North and South America (the Laramide Revolution); a fourth ushered in our own times and witnessed the birth of the Eurasian mountain chains (the Alpine Revolution). In the earlier stages of earth-history the evidence is less definite, but most authorities seem to agree that at least two great revolutions occurred in the Archaean and early Algonkian.

In what follows I shall endeavour to trace the origin and inter-relation of the events, referred to above, to: (1) The existence of a general basaltic magma-ocean or isostatic layer in which the continents float and upon which the oceans rest; (2) The presence of a certain amount of radioactive materials throughout this magma-ocean; (3) The maintenance, both in the past and in the present, of isostatic equilibrium of the land-masses; (4) Certain forces acting on the earth's surface crust—forces of astronomical origin, i. e. tidal and precessional.

The primary units of Geological Science are the same as those of Physical Science: Matter, Energy, and Time. This truism has long been recognized. But it is only recently that energy as the output of matter has entered into geological concepts.

Although the general distribution in the rocks of the energy-producing elements was demonstrated by the present Lord Rayleigh seventeen years ago, it is remarkable that geologists have in most cases paid but little heed to them. I say it is remarkable, *for the discovery carries with it the fact that we have everywhere in the rocks a perpetual source of heat; unfailing—whether the heat is accumulating or whether it is escaping—from age to age.* And we know of no physical conditions likely to modify this outflow of energy. Indeed, we may be sure that, far into the earth's interior such hypothetical modifying conditions do not exist. For, in fact, the source of the energy resides in the nucleus of the atom; that region which no effort of the chemist or physicist has ever reached save by the intermediary of the alpha ray.

That the output of radioactive energy dates back into the most remote periods of geological history is shown by the pleochroic haloes of Early Archaean rocks. Further, the law of radioactive decay renders it certain that more and not less thermal energy must have been evolved in those long-past ages. Certain features of uranium haloes seem to strengthen this view.

How then are we to deal with the consequences that must arise? Do we nowhere in the surface-changes of the earth see any evidence of the effects of the accumulated energy? I would answer that this accumulated energy has been a dominating factor in the history of the earth's surface-changes throughout the whole of geological time.

I have already referred to Lord Rayleigh's discovery of radioactivity as of world-wide significance. Along with this great advance in recent knowledge I class the discovery of isostasy, which, although dating its origin back to the middle of the last century, only recently became established; mainly in the work of Hayford and Bowie in the United States, and of Burrard in India.

Isostasy connotes the existence on the earth's surface of a layer of dense, plastic material in which the continental masses of lighter substances float, and upon which the oceans rest. And, just as large and heavy vessels are necessarily of deeper draught than lighter craft, so the more elevated continental features are 'compensated' by a more considerable displacement of the underlying sustaining layer. If compensation were perfect all over the world it would follow that vertical prisms, based on the floor of the isostatic layer and of equal horizontal section, taken anywhere on the surface of the globe would contain equal mass. It has been determined that this statement is only true if such prisms be taken of considerable horizontal dimensions; about that of a square degree upon the equator. In other words, the lesser features are not separately compensated. The rigidity of the crust distributes the load so that the compensation of the smaller features becomes merged in that of the land-mass as a whole. But the greater mountain ranges are more or less perfectly compensated by great downward extensions of the lighter surface materials into the sustaining substance. Thus the Himalayas are stated to be about 80 per cent. compensated. The depth to which such protuberances extend must be very great, for there can be only small difference of density between the submerged continental rocks and the supporting medium.

This theory has passed the stage of speculation. The measurements of both the direction and magnitude of gravitational force made in many parts of the world—notably in North America and in India—establish it.

Now the very interesting question arises as to the nature of the material forming the isostatic layer. The evidence that this layer is composed of basalt or basaltic-magma is very strong; I should say it is conclusive. The facts are as follows:

As referred to above, at intervals in geological history there have been great vertical movements of the continents, both in part, and apparently as a whole. We can only say of these movements that they are relative to the ocean; but that is sufficient for our argument. For it is impossible that continents or oceans can be relatively displaced vertically by one or more hundred fathoms unless the sustaining medium is, for the time being, in a fluid state, and capable of extensive lateral displacement. Attending these disturbances we find, consistently, that vast amounts—even to hundreds of thousands of cubic miles—of intensely heated, fluid basalt came up through fissures from beneath and flooded the land surface. The evidence is of the sort that would convince the occupants of a ship, when water flowed in at casual leaks, that their vessel must be floating in water. But the evidence offered to the geologist is of yet more convincing nature. For the continents are not hollow vessels but solid rock-masses. It must be true, therefore, that the buoyant substance is of greater density than the continental rocks. The substance basalt, even when fluid, possesses this property; Day, Sosman, and Hostetter have shown that quartzite or feldspathic sandstone floats on liquid diabase. I have repeated this experiment, using rock-crystal and liquid basalt. And there is yet another criterion. If the fluid is a melted rock, then its melting-point must be lower than that of the continental rocks. Here again basalt amply meets the requirements. This substance therefore possesses all the needful properties. Its enormous abundance is testified not only by the great lava-floods since the remotest periods of geological history, but by the fact that it is by far the most prevalent effusive rock on the globe. So also the oceanic islands are predominantly basaltic; and gravitational measurements support the view that this substance directly underlies the waters of the earth. In the opinion of many petrologists basalt is the parent rock-magma from which the several rock-families have been derived.

There appears to be no reason for regarding the continents as anything more than a highly silicated differentiate which has separated out and risen to the surface of a vast basaltic ocean some seventy or more miles in depth. This granitic scum has collected into those patches which we designate as Eurasia, Africa, &c. Upon the scum-free areas the oceans rest.

According to this view—and it is difficult either to replace or to assail it—the relation of the continents and the oceans to the sustaining basaltic layer are, in a sense, hydrostatically different. The continents float in the true Archimedian sense, by displacement of the basaltic substance. The oceans rest upon its surface as oil would rest on the surface of water contained in a vessel of limited area.

Now we have very strong reasons for believing that the isostatic layer cannot at the present time be fluid. Astronomers tell us that a fluid layer beneath the outer crust would have the effect of almost obliterating the observed oceanic tides. For the tidal wave raised in such a layer would carry the whole outer crust up and down, and vertical movements of the water relative to the land would be extremely small.

Again, horizontal tide-generating forces arising under such conditions would act so as to retard the west-to-east motion of the earth's crust, and so would tend to cause slipping of the outer crust of the earth over the underlying globe, and to act as a brake on the earth as a whole, thus increasing the length of the day. If such an effect had prevailed in historical times it should be revealed in the records of ancient solar eclipses. Examination of such records seems to show that any such increase in the day is almost insensibly small.

Not less important is the evidence derived from seismology. Primary and secondary seismic waves emerging at an angular distance of 10 degrees from the epicentre are distinctly separated. Such waves have penetrated to a depth of some 100 kms. (65 miles).

The secondary waves are mainly distortional in character, that is, they are such as cannot be transmitted in a liquid medium. Both Oldham and Knott therefore conclude that beneath the outer heterogeneous layer, about 18½ miles in depth, there exists a layer which consists of a uniform homogeneous substance which can transmit both condensational and distortional waves and in which the velocity varies with the depth. Similar physical characters prevail far into the earth's outer regions.

On the whole we must conclude that, while slow viscous movements in the substratum may take place, satisfying the conditions of isostasy, it behaves as an elastic solid towards rapidly changing stresses.

The past and the present conditions of the basaltic layer would seem, therefore, to be thermally and physically different. For, as we have seen, the conditions prevailing during certain periods of geological history *must* involve more or less complete fluidity of the sustaining substance. Now the simplest conclusion we can come to—and the most probable in that it involves the minimal thermal change—is that at the present time the basaltic magma is in the viscous-solid state very near its melting-point. In the past it was in a state of fluidity and still near its melting-point.

Let us, then, consider the existing state of the magma ocean as that of a viscous solid near its melting temperature, and from this as a starting-point, look forward into the future.

Accepting the view that the isostatic layer is composed of basalt, such as has been many times poured out upon the earth's surface, we know that it contains radioactive elements, and we possess, in fact, considerable observational evidence enabling us to form an estimate of the amount of heat continually being evolved thereby.

Now a substance in the solid state can only part with heat by the slow process of thermal conductivity supposing it abuts upon a body colder than itself. It will presently appear that *this condition involves the fact that throughout the great body of the basalt heat must be now accumulating. For the heat has no means of escape and the thermal evolution is unceasing.*

It is true that the rate of supply is slow—perhaps some 3 or 4 calories per grm. of the magma in one million years. But, looking forward some 25 or 30 million years, 90 or 100 calories will have accumulated. This involves liquefaction, for this is the amount of the latent heat of basalt.

Hence in some 30 million years the basaltic magma which now is solid will have changed its state, its temperature remaining much as at present. This time estimate is based on the mean value of the radioactivity of basalts. If the lowest values (those for the Hebridean basalt) were taken as applying to the whole magma, the period required for liquefaction would be raised to about 40 million years.[1]

How will the change of state affect the land and water surfaces of the globe? The voluminal expansion of basalt attending fusion, according to the admirable experiments of Day, Sosman, and Hostetter, appears to be some eleven per cent. of the volume of the unmelted glass or twelve per cent. of the volume of the crystallized basalt at the melting-point (1,150° C.). As fusion progresses the magma-ocean expands upwards, carrying with it both oceans and continents. But, as I have stated, the relation of the continents to the magma is not the same, hydrostatically, as that of the oceans. The oceans are borne upwards on the rising magma, although their surface must sink a little owing to increased area. Although the continents also are borne upwards they must sink in the magma owing to the lesser density of the fluid basalt. They experience on the whole a downward displacement relative to the oceans, continental regions supported by deep 'compensations' being most affected in this manner. This comes about very gradually as melting becomes more and more general throughout the magma. Moreover, local, more advanced, fusion beneath the continents will lead to surface movements long before the general fusion is completed.

Such surface changes of level must give rise to transgressional seas. These may creep in a little way as determined by local fluctuations of level, and then go on augmenting for long ages as the melting approaches completion.

We see that at this stage in our prophetic history of the earth we have attained conditions such as attend the earlier eras preceding a 'revolution': and, in fact, that a climax has been reached; in so far that the magma ocean is now entirely, or for the greater part, in a state of fusion. We shall consider later the orogenic effects which must ensue, and shall now continue to trace the thermal events after liquefaction is completed.

The former conditions of heat storage no longer prevail, for we deal with a liquid and not with a solid. In liquids, circulation and thermal convection take the place of conductivity and the rate of loss is enormously increased.

It is chiefly beneath the oceans that this loss will take place, and that the thermal accumulation of some thirty millions of years will, for the greater part, be discharged. During the period of heat storage the isotherm representing the melting-point of basalt

[1] These time estimates are less than those arrived at in the present volume (see p. 153 *ante*). The difference is due to the fact that the more recent estimates, in addition to the latent heat, take into account the fall in temperature below the melting-point which the basalt experienced when last it solidified, as inferred from the experiments of Day and his colleagues.

may have sunk to a depth of some 15 or 20 miles. The sub-oceanic crust so formed is now rapidly thinned by circulation of lava currents—doubtless often superheated—from beneath. Part of it may founder although its break-up on any considerable scale is improbable. Some 6 to 12 millions of years may be required for the heat to escape through the reduced ocean floor, according as we assume it reduced to $\frac{1}{4}$ or $\frac{1}{2}$ its original thickness. The congealed magma gravitates downward and the magma stratum solidifies from the depths upwards. This heat-loss gradually ushers in the decline of the revolution and the commencement of a new era of thermal accumulation.

We must now again revert to the surface effects attending these changes. With the increased solidification of the magma everything is reversed. There is general sinking of the earth's surface, both continents and oceans gravitating downwards owing to the contraction of the magma. Again, however, there is a differential effect, for the continents emerge in obedience to the increased isostatic forces arising out of the augmented density and thus the transgressional seas must, little by little, vanish from the land.

When, instead of looking forward into the future, we look back into the past, we find all the events of this history written upon the face of the continents; perhaps most beautifully on that of North America and, pre-eminently, from Jurassic times onwards.

In this matter we have as much right to look back into the past as forward into the future; for the whole sequence appears to be inevitable, although we are not in a position to fill in the exact measurements, whether in space or in time. We cannot know—we probably shall never know—how much heat may be supplied from radioactive sources still deeper than the isostatic layer. We do not know the depth of this layer, nor can we evaluate the physical conditions affecting its great depths. We cannot lay out the timing of these cyclical events with any claim to accuracy. We can only say of such time-periods as seem probable that the recorded great world-revolutions might have been consummated in the lesser estimates of geological time. A total of about 40 millions of years, possibly 50 millions, may be involved in the consummation of the physical events responsible for a world-revolution.

As I have said, the conditions of cooling of the magma are mainly operative beneath the oceans. The continents offer a nearly adiatherminous covering to the magma. The reason for this is not far to seek. The continental rocks are considerably more radioactive than the basaltic layer. If, on the evidence of seismic phenomena, we assume for the continents a mean thickness of some 32 km. (20 miles), it is easy to show that the intrinsic radioactivity of the more acid rocks gives rise to a basal continental temperature the same as, or but little less than, that of the melting basalt, and approximately accounts for the mean continental gradient of temperature as observed at the surface; in other words, for almost the whole heat escaping over the land-areas of the earth. But little, therefore, can be referred to radioactive heat generated in the underlying magma. In fact, calculation shows that the thicker parts of the continental layer must transmit a part of their radioactive heat downwards into the magma. If, then, there were no horizontal circulatory movements in the fluid basaltic layer during times of revolution, the sub-continental temperature must rise indefinitely. It seems certain that, locally, superheating actually occurs, and that batholithic invasions of granitic magma into the rising mountain masses may be accounted for in this manner.

Certain tidal effects, which are to be expected as the result of conditions prevailing during times of revolution, adequately account for the equalization of temperature throughout the basaltic substratum. Darwin long ago pointed out that the horizontal tide-generating force of moon and sun, acting most effectively on the outer parts of the earth's surface, must tend to withhold the outer crust from partaking of the full west-to-east rotational velocity of the body of the earth. So that, if a fluid layer intervened between the outer crust and the underlying core, a differential motion must arise. The result is at the present time negligibly small—as already referred to—the tidal force not being sufficient to shear the solid substratum. But in times when this substratum is replaced by a highly mobile fluid the displacement is inevitable. Thus the continental masses would slowly replace the oceans in position respecting the deep-lying magma. Or, in other words, the floor of the oceans must come to overlie magma which has probably been superheated beneath the continents during the long period required for thermal accumulation. These conditions, leading at once to rapid liquefaction and thinning of the sub-oceanic floor and the escape of sub-continental heat, arise from the astronomical forces to which the earth is exposed.

The subject of the circulatory movements which arise in the fluid magma during times of revolution will now be further considered in connexion with the orogenic events attending great revolutions.

We have hitherto considered mainly the thermal changes and the vertical movements of the earth's surface which attend these thermal changes. But along with the vertical movements of the surface certain horizontal forces come into operation which conspire with the vertical movements to bring about mountain-building on the land-areas. And here I venture to offer some suggestions as to the nature of orogenic movements, and more especially with regard to the magnitude of the forces involved.

In current views of the mechanism of mountain elevation, the part played by horizontal forces has, in my opinion, been over-estimated and that of vertical forces under-estimated. I notice in a recent paper by a high authority on isostatic theory—W. Bowie—that this point is emphasized.

That mountains arise in geosynclinal areas is well known. But what is the nature of these areas? They are elongated troughs in which sediments have collected throughout long periods of time. In general they range parallel with a sea coast. This may be because their location has been determined by prior-existing crustal elevations—possibly of small dimensions—which have arisen along the continental margin in a manner to which I shall presently refer; or, more rarely, because the sediments have been actually laid down in a shelf, or coastal, sea.

The sedimentary collection in the geosyncline may be some miles in depth and still go on augmenting. We know, therefore, that in obedience to isostasy it is sinking, and that the underlying continental mass is being pressed into the highly heated isostatic layer beneath. We have to remember that while this process is going on, and transgressional seas are extending, the magma is melting down into the fluid state, and is approximating more and more to the density of the continental rocks. Hence when, later on, horizontal orogenic forces come into action along the continental margin, they operate on a trough—perhaps 1,000 miles long or more—loaded with loosely consolidated sediments resting on a highly heated and much bowed part of the continental crust. Their effect is not to push up a mountain, but mainly to urge the compressed layers deeper into the molten magma. For, in fact, any other result is opposed to isostatic conditions. It is at this time that the folding and over-thrusting, &c., are for the greater part brought about. The mountains are not raised till long afterwards. They arise when the period of fluidity has run its course and given place to that of advancing solidification. The density of the isostatic magma is then rising. The compensation becomes excessive and the whole crushed mass is pushed upwards. This is not due to the horizontal but to the vertical forces. The energy comes directly from the potential energy of the expanded magma, and is, of course, traceable to the radioactive heat accumulated in long-past ages.

We see, then, that we are not called upon to adduce horizontal forces of enormous magnitude, adequate to force upwards the ranges of the Himalayas or the massif of Mt. Blanc, but rather such as will supply the comparatively small pressures required to compress laterally the heated materials of the geosyncline as these, almost automatically, subside into the fluid magma beneath. It is with these views of mountain genesis in mind that we should turn to the consideration of the origination of orogenic forces.

I have already made a passing reference to the horizontal tidal force of which astronomers tell us. During the period of fluidity of the isostatic layer, an ocean, probably not less than seventy miles in depth, and possibly very much more, comes into existence; an ocean having a density three times that of water. The tidal movements generated in this ocean must greatly exceed in energy those of the ocean of water, and must, in like manner, produce a retardative effect on the west-to-east movement of the earth's surface and cause a relative motion of the outer crust over the inner parts of the earth.

Now it is not hard to see that, if the continents and ocean floor are retarded by the horizontal tide-generating force from partaking of the full angular velocity of the globe, the magma-ocean must exert an easterly directed force on the submerged continental features. For the lowest layers of the magma must, in virtue of their viscosity, more nearly preserve the full angular velocity of the earth than the upper parts. In this way there must arise a west-to-east pressure transmitted from the rotating earth in opposition to the lunar and solar gravitational forces.

But this is not the only force of astronomical origin which will act in this way. W. Schweydar finds that a precessional force acts upon the continents (in the same direction as the tidal force) but of much greater intensity than the latter. It is a maximum at the equator and is nil at 36° north and south of the equator.

The horizontal magmatic pressure called into existence by these forces can only act on the westerly submerged coasts of the continents and on the westerly side of downward-reaching obstructions. Now it is a fact that the great lava-flows of the past were poured out mainly on the westerly sides of the continents or on the west of mountain ranges, e. g. the Hebridean, the Deccan, the flows of Western North America (many of them inter-montane). Then there are minor flows, such as the Disco flow west of the Greenland ranges; and the volcanic and effusive phenomena arising along great north and south rifts, e. g. those of the African continent.

As for the orogenic effectiveness of these magmatic pressures of astronomical origin we should, as I have stated above, bear in mind what the orogenic force is required to perform. However, until their mathematical evaluation is accomplished—if this is indeed possible with the available data—it is, of course, premature to discuss quantitatively effects about which we only can say that they must be far greater than such oceanic tidal forces as now affect the earth's surface and expend their energy in land-locked basins. It is certain, too, according to lunar theory, that in the past they would be augmented by a closer approximation of the moon; the horizontal tidal force due to lunar attraction increasing as the inverse sixth power of the lunar distance. One other remark may be added: the effects of the entire horizontal gravitational pull on the tidal protuberance must be considered as concentrated over the limited areas of such obstructing features as catch the east-going pressure of the magma.

An outstanding topographic feature of the earth, the grandeur of the Cordilleras of North and South America as contrasted with the orographic features of the western side of the Pacific, may possibly have found origin in the effectiveness of forces of astronomical origin in conjunction with effects now to be considered.

There are other orogenic forces arising out of the present views which suggest possibilities of considerable importance. They arise from the thermal expansion and subsequent contraction of the magma-ocean when changing its physical state.

If we assume that the average volume-change of the magma ocean is from all causes about 10 per cent., we find that this involves an increase of the earth's radius of 6·5 miles, assuming that the isostatic layer is 70 miles deep. The surface area of the globe increases correspondingly by 650,000 square miles. Practically all this increase falls upon the oceans because the continents are, and remain, essentially homogeneous. Hence we find that the equatorial width of the Pacific must increase by about 30 miles and that of the Atlantic by 11 miles. Now these increases are not very great, but it is quite possible, some authorities would say highly probable, that the isostatic layer is much more than 70 miles deep. If we take the layer as 100 miles in depth, the increase of the earth's surface would, very approximately, vary proportionately.

It is desirable to arrive at some idea of the thickness of the oceanic floor—or sub-oceanic crust—which is concerned with these movements. This floor, as already defined, is that surface layer of the magma which directly supports the ocean and which loses heat to the ocean during the long period of thermal accumulation Its downward extension may be considered as defined by that geotherm which attains approximately the general melting temperature of the underlying magma.

The problem of estimating its thickness is, to a first approximation, similar to that investigated by Lord Kelvin when estimating the downward distribution of temperature in a globe cooling from fusion; when the heat escapes by conductivity into a surrounding medium maintained at a constant lower temperature. Kelvin's value for the diffusivity is not applicable because taken for unsuitable rocks. Again, some part of the data which Kelvin asked for is now available. The mean specific heat of basalt between 0° to 1100° is known from the experiments of Barus. The variation of its thermal conductivity at high temperature is not known. Daly states that it decreases rapidly with rise of temperature. H. Poole, dealing with the crystalline rock, finds that up to 600° C. there is no change. Neglecting the change in conductivity, the change in specific heat alone involves a large decrease in the value of the diffusivity.

Using the amended value for this constant, it appears as a first approximation that in 25 millions of years a crust about 20 miles thick would form, at the base of which the temperature would approximate to that of the melting-point of the basalt. But this is excessive, for the effects of the intrinsic radioactivity of the cooling layer is not here taken into account. An allowance for this would leave the crust forming in inter-revolutionary times as about 15 miles in thickness. This crust is what I have called the *ocean floor*, and is distinguished from the underlying magma, which may also be solid, in possessing a definite gradient of temperature.[1]

During the period of thermal dissipation the ocean floor will be attacked by hot and, doubtless, often superheated currents from beneath, and rapidly reduced in thickness. Under the stresses arising out of the increased surface area of the ocean and the vertical displacement of the continents, acting together, it will rupture around the coasts. It is possible that magma currents, already referred to, may result in locating such fractures principally on western coasts. The fractures will be rapidly filled in by congealing basalt. It is not improbable that the location of the Hebridean lava-flow was determined in this manner. It is important to remember that during the period of fusion the earth's crust is stressed by forces that resist the expansion of the magma. *The magma is therefore in a state of pressure.*

The struggle between the sub-oceanic crust and the thermal forces beneath may endure for a considerable period—probably from 5 to 10 millions of years. During this long period the whole or greater part of the latent heat of the magma has passed into the ocean (at no time sensibly raising its temperature) and the congealing of the basalt has progressed from beneath upwards throughout the isostatic layer. Towards the end of it the ocean floor is being thickened and strengthened.

But the effect of the volume-change of the basalt, as we have seen, is to cause the down-sinking of the entire magma-surface and the return of the oceans to their original areal extension. The enlarged crust, forced into a more restricted area, must, on every side, bear upon the continents. The surface crust previously exposed to tensile forces is now compressed.

It seems certain that this, in fact, is the source of those forces which do the major lateral orogenic work. It persists for such a period as suffices to dissipate the energy and relieve the major stresses ; and it is evidently capable of overcoming any exceptional resistances which might arise. Possibly the buckling of the

[1] See *Phil. Mag.*, vol. xiv, June 1923, p. 1175.

eastern and western margins of the Pacific floor are survivals of these effects.

All these activities are recurrent in character, acting in the same way in successive revolutions. Thus the Rocky Mountain Ranges (which have been estimated to represent a crustal contraction of 25 miles) are the product of two great Revolutions, the Laramide and the Cascadian.

On this view of the origin of lateral orogenic stresses we are reminded of that pregnant saying of Dana's that the loftiest mountains confront the widest oceans. Nowhere around the Atlantic are such mountains formed as confront the Pacific. The east-to-west ranges of Eurasia effectively confront the widest oceanic stretch upon the globe.

The source of those crustal disturbances which affect the continents during inter-revolutionary times I have only incidentally referred to in the foregoing remarks. I do not think there is any special difficulty raised by local movements even on a considerable scale—such movements, for instance, as disturbed the continent of North America throughout Jurassic and early Cretaceous times; or, again, those which followed upon the Laramide Revolution. The first are, I think, to be regarded as preparatory; the second as sequential, in origin.

The preparatory period must witness the local breakdown of the isostatic layer from solid to liquid, and at a later date the effects, possibly, of superheating. Earth movements and volcanism must result. And it is difficult to disconnect vertical from horizontal movements when the physical properties of rock-materials towards tensional and shearing stresses have to be considered. For these stresses produce fracture, and an ever-present sub-continental layer of molten rock under great pressure supplies materials which fill every dike; rendering return to the original dimensions impossible. Hence vertical oscillatory movements become a source of ever-extending lateral pressures.

The post-revolutionary period, on the other hand, witnesses the dissipation or relief of accumulated strains and the re-establishment of isostatic equilibrium. And these are sources of movement which may preserve their potentiality over considerable periods, for only in crustal movements can they find relief and exhaustion. Such relations, for instance, between the continental margin and the ocean floor as we have already referred to, might for long prevail—awaiting, as it were, the effects of denudative demolition or denudative accretion for their relief.

A very fundamental matter seems to find explanation in the foregoing considerations. I refer to the quantitative distribution of land

and water upon the globe. What has determined the relative areas of land and water upon the earth's surface? Might not the ocean have been deeper or shallower and the land area correspondingly increased or diminished? In other words, what has determined the average depth of the ocean and the average thickness of the floating continental crust? A simple answer seems to be forthcoming in the temperature conditions which calculation shows must affect the base of the continental crust owing to its intrinsic radioactive content. It is easily shown that in the thicker parts of the acidic crust—the Tibetan Plateau for example—heat must flow downwards into the magmatic substratum as well as upward to the surface. And this condition must prevail in many parts of the crust where its depth is as great as 35 km. if, in agreement with general opinion, we assume it to be mainly granitic in character.

During the long period of thermal accumulation, such supplies from above must be attended with superheating of the sustaining magma, and as there is no relief save in times of Revolution, it is possible that melting of the continental rocks themselves may locally occur. Material so liquefied can only escape upwards (forming stocks or batholiths) or laterally. The first alternative, which seems to have specially affected the Archaean crust, offers no permanent cessation of the phenomenon. Lateral escape alone can restore ultimate thermal equilibrium. This lateral escape must take place when tidal effects, already referred to, shift the entire surface crust. Fused acidic materials would then migrate from beneath the continental layer and enlarge its surface extension. Such actions, extensively affecting the early tectonics of the earth's crust, must very evidently determine not only the final area of the land but also the extension and depth of the ocean. To-day such activities probably limit the heights of mountain ranges and of the great plateaux. For in fact they define the efficacy of isostasy as the determining factor in sustaining the greater raised features of the globe. A crust thicker than what seismic phenomena indicate, if formed in earlier times—either fortuitously or under surface forces arising out of the earth's axial rotation—would not be stable. It must spread out upon the surface of the magma until it attained its present thickness and areal extension.

The thesis that the continents may have shifted during geological time gains some support from the foregoing history of terrestrial tectonics. For we observe that in the course of this history the ocean floor has been repeatedly assailed by thermal effects tending to reduce its thickness and rigidity, and that the continents were exposed to long-continued magmatic forces acting from west to east. If the ocean floor were in times of Revolution so far reduced in rigidity as to yield to those forces continental movements in an easterly direction would probably have taken place. We cannot now evaluate the intensity attained by the several factors concerned, and so we must leave to other sources of evidence the final verdict in this interesting theory. An affirmative answer would not be out of harmony with possibilities arising out of the present views.

It may be suggested that some other issue than those which have been outlined above might have been evolved from the conditions. That some sort of equilibrium between thermal gains and losses might have come about, and a steady flux of heat to the surface have replaced the cyclical phenomena, which are herein ascribed to radioactive energy. However, very grave difficulties arise when the assumed state of equilibrium is investigated. To instance one such difficulty. Without a periodic means of thermal escape to the surface, the temperature beneath the continents mounting up since pre-Cambrian time must now have arisen to much over 1,000° above the melting temperature of the continental granites and gneisses. Nor is it possible to explain the past without the recognition of periods of fluidity affecting the substratum.

Without any violence done to our logic we may appeal to the verdict of geological history in this matter. A theory of equilibrium, even were it possible to frame such a theory, leaves the whole history of the surface changes of the globe without a hope of explanation. Such cyclical changes as are outlined above not only arise consistently and naturally—we may say inevitably—from the conditions—conditions recognized by some of the foremost geological thinkers—but they faithfully explain many of the great facts of earthly tectonics. The events of the past cease to be mysterious. They become the natural outcome of the physical structure of the earth's surface.

# GLOSSARY

ANORTHOSITE. A rock mainly composed of a feldspar rich in calcium and relatively poor in silica. Like the granites, it crystallized slowly and under plutonic (or deep-seated) conditions. Enormous laccoliths composed of anorthosite enter into the structure of certain mountain chains, notably in Western North America. This rock contains less silica than the granites but more than the basalts.

AQUEO-IGNEOUS. A descriptive term applied to the mode of origin of certain minerals and rocks, the formation of which from a molten state is believed to have been influenced by the presence of water dissolved or intermixed in the magma.

AUGITE. A principal constituent mineral of basalt, gabbro, and dolerite. It is a silicate of alumina, lime, magnesia, and iron.

BASALT. A 'basic' rock generally containing about 50 per cent. of silica and about 12 per cent. of iron oxides, hence its high specific gravity contrasted with the granites (q.v.). The basalts are effusive rocks or, if injected, have cooled rapidly. Sometimes a certain amount of uncrystallized material (glass) remains. Basalts are the effusive representatives of the gabbros and dolerites. They are by far the most abundant effusive rocks on the face of the globe.

COMPENSATION. That part of the surface crust which, owing to its downward penetration into the heavy substratum, displaces such a mass of this as suffices to float some superincumbent surface feature. This displacement of the heavy substratum, on Archimedes' principle, just 'compensates' gravitationally the mass of the raised surface feature. The principle of isostasy is contained in this statement when applied over the whole surface of the globe.

DEEPS. Oceanic areas of exceptional depth, representing depressions in the ocean floor, often trough-like or synclinal in form. The term is generally understood to be restricted to depths greater than 18,000 feet (3,000 fathoms).

DIABASE. A name applied to dolerites which have undergone a certain amount of internal alteration in the course of time and hence generally refers to ancient doleritic rocks.

DIASTROPHISM. A term proposed by Powell to denote all the processes of deformation of the earth's crust. Elevations, subsidences, folding, faulting are all 'diastrophic'. Orogenesis and epeirogenesis (q. v.) are forms of diastrophism. (See Geikie, *Geology*, i, p. 392.)

DIFFUSIVITY. Thermal conductivity ($k$) has been defined in the text (p. 72). Diffusivity is the thermal conductivity divided by the thermal capacity per unit volume, i. e. by the quantity of heat in calories required to raise 1 c.c. of the substance one degree in temperature.

Diffusivity, i. e. $\frac{K}{C}$, is analogous to the 'coefficient of diffusion' required in calculations dealing with the quantity of dissolved substance passing through unit thickness of the solvent in unit time.

DIKE or DYKE. A fissure extending more or less vertically and filled with an igneous injected rock.

DOLERITE. This rock is formed when a basaltic lava is injected into veins or dykes so as to cool under some pressure and slowly. It is completely crystallized. It is the same in chemical and mineral composition as a basalt, but is more coarsely crystallized.

DYNE. The unit of force. It is that force which acting for one second upon the mass of one gram would confer upon it a velocity of one centimetre per second, or, if acting continuously, an acceleration of one centimetre per second per second. As the weight force (due to gravity) confers upon a falling body a velocity of 981 cm. per second (in London or Paris), the weight of one gram is a force of 981 dynes. Thus the weight of a milligram is a force a little less than one dyne.

EPEIROGENIC. A term applied to forces and movements leading to continental development.

EPICENTRE. The point on the earth's surface vertically over a centre of seismic disturbance.

FAULT. Faults are fractures of the earth's crust, which may extend for hundreds of miles, and in which there has been a relative displacement of the crust at opposite sides of the fissure. If the plane of the fracture is inclined to the vertical, and the displacement has taken place so that the overlying rock has been displaced downwards over the underlying part, the fault is said to be 'normal'. If the overlying side has been displaced upwards the fault is said to be 'reversed'.

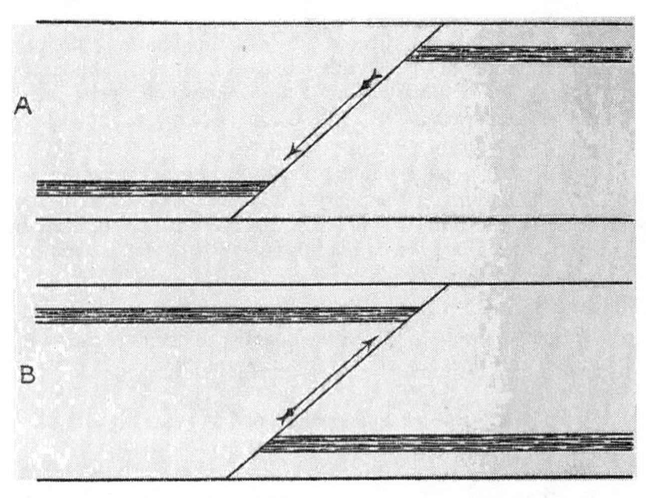


FELDSPAR. A very abundant family of minerals common in all igneous rocks except in the ultra-basic group from which they are almost or entirely absent. Feldspars contain in all cases silica and alumina. The other constituents may be potash, soda, or lime. The quantity of silica varies with the silica-richness of the containing rock; the feldspars in granites containing a great deal of silica; those in basalts considerably less.

FLUORITE OR FLUORSPAR. This mineral is a fluoride of calcium. It is not one of the rock-forming minerals, and is found sporadically in connexion with ores of lead, &c. Lately very perfectly developed haloes derived from uranium have been found in this mineral.

GABBRO. This rock is the plutonic or deep-seated representative of the basalts. The gabbros are similar to basalts in chemical composition but are always completely and coarsely crystallized.

GEANTICLINE. A bowed-up, or arched, deformation of the surface crust extending over a large area.

GEOSYNCLINE. A great trough-shaped depression in the surface crust which becomes filled by transgressional waters and for long ages collects detrital sediments.

GEOTHERM. Or isogeotherm; an imaginary plane beneath the surface of the earth at every point of which the temperature is the same: a plane of uniform temperature.

GNEISS. The gneisses are banded and foliated rocks which have experienced considerable dynamic metamorphism, i. e. have been exposed to great pressures under which shearing motions have taken place. Very often they are granitic in composition, the constituent minerals being segregated in layers. They are completely and visibly crystalline throughout. See Mica Schists.

GRANITE. The granites consist essentiallyof quartz, feldspar, and mica. They contain a large percentage of silica and are accordingly described as 'acid' rocks, silica being an acid-forming substance in presence of water. They are plutonic (q. v.) in origin and completely, and generally coarsely, crystallized. They are, as a group, the most radioactive rocks on the surface of the earth. They form great laccoliths (q. v.), which seem in some cases to enter into mountain structure as a core which has been forced upwards from beneath by great pressures.

HALF PERIOD. A term used in the science of radioactivity to denote the period required for one half the present amount of a substance to be radioactively transformed. Although the substance is continually decreasing in quantity, the value of the half period remains the same, provided the substance is radioactively homogeneous. In which case the diminution of the substance evidently proceeds in geometric progression, the quantities remaining in successive half periods becoming $\frac{1}{2}$, $\frac{1}{4}$, $\frac{1}{8}$, $\frac{1}{16}$, &c., of the original amount. If, however, the observed rate of decay has been derived from that of a constituent isotope (see p. 150), then the rate of decay (i. e. the time value of the half period) will vary with the lapse of time; growing greater by degrees; for the radioactive constituent present is continually becoming a less and less fraction of the whole.

LACCOLITH. An intrusive mass of igneous rock—often of very great dimensions—generally associated tectonically with a mountain range and elongated parallel with the orogenic axis. It is domed above and at least in part below, and is supposed to have been forced by great pressure between the bedding planes of the enclosing rocks.

MAGMA. This term is generally applied to the primitive molten lava from which the various rock clans are believed to be derived, or the term may be applied to any molten rock-making material. In the text it is applied to the general mass of the highly heated substratum, whether in the solid or in the fluid state.

MAGMA-BASALT. Applied to a class of basalt in which the crystallization has been imperfect, much glass being present. It is also specifically applied to certain ultra-basic rocks (q. v.).

MICA. One of the three distinctive constituents of granite and abundant in mica schists and many gneisses. It is remarkable for its very perfect cleavage. There are two varieties: Muscovite, which is silvery white and is a hydrated silicate of alumina and potash; biotite, which is brown in colour and is a hydrated silicate of alumina, iron, magnesia, and potash. The haloes due to uranium or to thorium are found in biotite, not in muscovite. It seems probable that their visibility is due to the iron present in biotite.

MICA SCHISTS. A schistose and foliated rock composed essentially of mica and granular quartz, the schistose character being referable to the mica flakes which are orientated in parallel planes. A metamorphic rock.

OROGENESIS. The development of mountains under orogenetic or mountain-making forces.

OVERTHRUST. When a bedded rock is folded vertically by the action of lateral forces, and subsequently exposed to further lateral movements acting more effectively towards the top of the fold, the whole fold becomes bent over horizontally. If the shearing movement continues, the upper half of the fold may be displaced over the lower half. This would be described as an overthrust. Various forms and degrees of similar movements are termed overthrusts. Such movements may extend under orogenetic forces for many miles, as in the case of the Alps.

PENEPLANE. An imperfectly developed plane (so, also, peninsula—an imperfect island). It is the final stage of mountain wastage and decay, e. g. the great Archaean Shield of North America.

PLUTONIC. Descriptive of a rock which has been formed deep down under conditions of high pressure and slow cooling. See Granite, Gabbro, Anorthosite.

RECUMBENT FOLDS. These are rock-folds which have been pressed over by unbalanced horizontal forces so directed as to cause the fold to take a horizontal or recumbent position. In the Alps such folds have been traced for scores of miles.

SILL. A fissure in the surface crust extending more or less horizontally; generally a lateral branch from a dike (q. v.) which has been injected with a molten rock.

TECTONIC. A descriptive term applied to features in the surface crust produced by deformative forces, e. g. mountains, valleys, faults, &c.

TRACHYTE. A volcanic rock containing less silica than the granites and more than the basalts. It is generally fairly completely crystallized and largely composed of minute feldspars.

ULTRA-BASIC. A descriptive term applied to certain rocks very poor in silica (less than 45 per cent.), containing no feldspar and much iron. They exceed the basalts in specific gravity and are not quite so radioactive.

CPSIA information can be obtained
at www.ICGtesting.com
Printed in the USA
LVHW022211170323
741893LV00032B/1343